Making a Living

Livelihoods in rural Africa are changing to respond to disappearing job prospects, falling agricultural output and collapsing infrastructure. Responses to these challenges are very different in different parts of Africa. This book explains why.

Making a Living explores the local histories and livelihood strategies underlying these diversities. It uses case studies from Eastern and Southern Africa to give a broad comparative study of rural livelihoods. Many people in rural areas have long depended on labour migration to urban areas to make a living. Urban jobs have become scarce. Many migrants cannot afford to send money back home, others are returning and trying to make a living locally. Many young people are not migrating in the first place. This book examines how people are responding to this major change in the political economy of rural Africa.

The case studies range from commercial farming regions in Kenya, Tanzania and Zimbabwe to much poorer areas in Kenya, Southern Zimbabwe, Northern Zambia and South Africa, where rural livelihoods have long been dominated by labour migration and where multiple livelihoods are increasingly common. *Making a Living* also gives a detailed exploration of people's lives and livelihoods in two localities, Kisumu District in Western Kenya and North West Province, South Africa.

Elizabeth Francis is a lecturer in Development Studies at the London School of Economics. She has previously taught Sociology at the University of Essex. She has carried out research on rural livelihoods in Kenya and South Africa and has published articles on rural livelihoods in *Africa, Journal of Development Studies, Journal of Contemporary African Studies* and *Canadian Journal of African Studies*.

Making a Living

Changing livelihoods in rural Africa

Elizabeth Francis

London and New York

First published 2000
by Routledge
11 New Fetter Lane, London EC4P 4EE

Simultaneously published in the USA and Canada
by Routledge
29 West 35th Street, New York, NY 10001

Routledge is an imprint of the Taylor & Francis Group

© 2000 Elizabeth Francis

The right of Elizabeth Francis to be identified as the Author of this Work
has been asserted by her in accordance with the Copyright, Designs and
Patents Act 1988

Typeset in Galliard by Taylor & Francis Books Ltd
Printed and bound in Great Britain by St Edmundsbury Press, Bury
St Edmunds, Suffolk

British Library Cataloguing in Publication Data
A catalogue record for this book is available from the British Library

Library of Congress Cataloging in Publication Data
Francis, Elizabeth
Making a living: changing livelihoods in rural Africa / Elizabeth Francis.
p. cm.
Includes bibliographical references and index.
1. Africa, East–Rural conditions. 2. Rural poor–Africa, East. 3. Africa,
Southern–Rural conditions. 4. Rural poor–Africa, Southern. I. Title.

HN792.A8 F73 2000-01-20
307.72'096–dc21
                                99-058782

ISBN 0–415–14495–7 (hbk)
ISBN 0–415–14496–5 (pbk)

For Gruffydd Francis
1928–1999

Contents

Tables and illustrations

Tables

Figures

Plates

The plate section follows p.98

Acknowledgements

Financial support for my Kenyan research came from the Economic and Social Research Council, the Wenner-Gren Foundation, the Royal Anthropological Institute, Somerville College, Oxford and St. Anne's College, Oxford. My research in South Africa has been funded by the Wingate Foundation, the Leverhulme Trust and the UK Department for International Development.

I am grateful to everyone who has given me help, advice and inspiration to write this book. In Kenya, I must thank Asenath, James and Adhiambo Odaga, who made my stay in Koguta possible, Patricia, Jacton and Esther Akumu, who took me into their family, and Ruth Waga, who worked as my research assistant. I also owe a great debt to the people of Koguta, for their welcome and tolerance. I should also like to thank Professors H.W.O. Okoth-Ogendo and John Oucho, of the Population Studies and Research Institute, University of Nairobi, for giving me an academic base in Kenya.

My research in South Africa has been made possible by the generosity of many people. James Drummond encouraged me to do research in North West. Staff at the North West Departments of Agriculture and Land Affairs gave practical support. Ben Mosiane and Nancy Moiloa were enthusiastic and insightful research assistants.

Molly and Gruffydd Francis supported me always. I owe much to Gavin Williams, Judith Heyer, Megan Vaughan, Parker Shipton, Teddy Brett and Colin Murray, who all taught me to think critically about social change in Africa, and to Terry O'Shaughnessy, who encouraged me to think critically about everything.

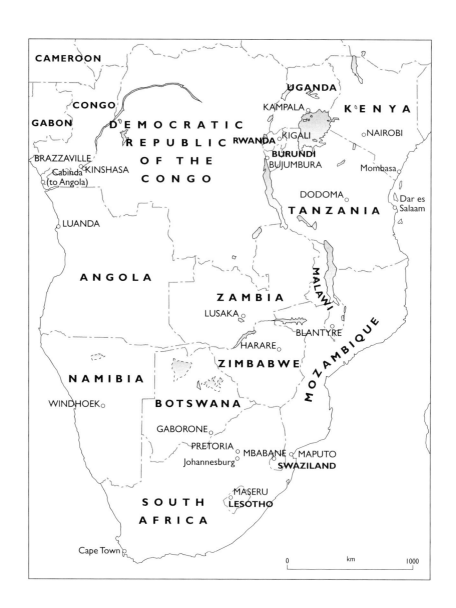

Southern and Eastern Africa

Introduction

This book is about the different ways in which people in rural Africa try to make a living. Its focus is on eastern and southern Africa, using case studies to understand the livelihoods of people who depend on various combinations of small-scale farming, livestock-keeping, trading and migrant labour. It looks at how people living in rural areas have responded to the falling agricultural output, disappearing job prospects, collapsing infrastructure and withdrawal of public services that have characterised economic change in these regions since the 1970s.

Economic decline, poverty and deprivation

Per capita income data give some indication of the scale of these problems. Table I shows data on real gross domestic product (GDP) per capita between 1960, the eve of Independence for eight of these countries, and 1992. Comparable data for some other developing countries is included. They emphasise how far Africa has fallen behind the rest of the developing world in the last forty years. Catastrophic falls in income per capita in Angola and Mozambique, as well as income decline in Uganda, can be explained by the effects of decades of war, which, in the case of Angola, show no prospect of ending. Mineral exports underpin rising per capita income in Botswana. Elsewhere, slightly rising incomes in the 1960s gave way to stagnation or decline in the 1970s and 1980s. Per capita income remained stagnant in Kenya between 1980 and 1992 and throughout the 1970s and 1980s in South Africa. In the same period it fell by 10 per cent in Malawi. Zambian real GDP per capita declined by almost 40 per cent between 1970 and 1990. Per capita income in Zimbabwe rose during the 1970s. It held stable in the early 1980s but later began to fall. Table II shows where these problems have persisted in the 1990s. In Lesotho, Uganda and Mozambique the picture is more encouraging, although economic growth in Uganda and Mozambique has been from a low base.

As well as stagnation and absolute decline in incomes, Sub-Saharan Africa has fallen back relative to the rest of the world. Countries whose GDP per capita was similar to Sub-Saharan Africa in 1960 have increased incomes, several fold in the case of the Newly Industrialised Countries of East Asia. Despite a rapidly increasing population, Indian GDP per capita rose by 67 per cent between 1960 and 1992. In China, per capita income in 1992 was over two and a half times its level in 1960. Angola, Zimbabwe and Zambia all had higher per capita incomes than South Korea in 1960. Taiwan's per capita income was similar to that of Mozambique and $935 less than that of South Africa.

Explanations of Africa's economic decline, as well as its failure to take part in the economic growth seen in other developing countries, must include the long-term impact of Africa's incorporation into the world economy during and since the colonial period; various, different, pathologies of state-society relations experienced in Sub-Saharan countries; the impact of decades of war in a large

Table I Real GDP per capita, 1960–92 (1985 $US prices)

	1960	1970	1980	1990	1992
Angola	931	1165	675	n.a.	n.a.
Botswana	535	823	1940	n.a.	n.a.
Kenya	659	586	911	911	914
Lesotho	313	419	994	972	952
Malawi	380	440	554	519	496
Mozambique	1153	1497	923	760	711
Namibia	1790	2642	2904	2854	2774
South Africa	2191	3254	3496	3248	3068
Tanzania	319	424	480	n.a.	n.a.
Uganda	598	647	534	554	547
Zambia	965	1117	971	689	n.a.
Zimbabwe	989	1082	1206	1182	1162
India	766	802	882	1264	1282
China	567	696	972	1324	1493
South Korea	904	1680	3093	6673	n.a.
Taiwan	1256	2188	4459	8063	n.a.

Source: Summers and Heston (1991), 'The Penn World Table', version 5.6

Note: The Penn World Tables have been developed in order to make meaningful cross-country comparisons of national accounts data. See Summers and Heston (1991) and http://www.pwt.econ.upenn.edu

Table II Per capita GDP growth, 1990–97

Angola	-2.4
Botswana	n.a.
Kenya	-0.6
Lesotho	5.5
Malawi	0.9
Mozambique	3.1
Namibia	1.5
South Africa	-0.2
Tanzania	n.a.
Uganda	4.1
Zambia	-3.3
Zimbabwe	-0.3

Source: World Bank, *World Development Report*, 1998/99

number of countries and the macro-economic impact of the Structural Adjustment Programmes so widely adopted (though less widely implemented) as remedies for Africa's economic problems in the 1980s and 1990s (Bayart *et al.* 1999; Bloom and Sachs 1998; Collier and Gunning 1999; Mamdani 1996).

Economic decline has drastic impacts on poverty and human welfare. Table III suggests how severe these effects have been.

All the countries have a high score for at least one of the indicators of poverty and deprivation, though there are some significant differences between countries in the values of the different indicators. Levels of extreme poverty, indicated by command of less than US$1 per day, appear to be lowest in Tanzania. A possible explanation for this finding comes from a study of rural responses to structural adjustment programmes in Tanzania carried out in the early 1990s (Booth *et al.* 1993). This research suggested that liberalisation of the economy was making it easier for households to diversify production, benefiting poorer households, as well as the better-off. However, the 27 per cent of children under age five who are underweight suggest that many Tanzanians live in severe poverty. The report that 35 per cent of the population of Botswana earn less than US$1 per day confirms the skewed distribution of income in that mineral-rich country. The percentages of the population not expected to reach the age of forty are extremely high compared with other developing countries. The comparable figures for India and Bangladesh are 16.1 per cent and 21.5 per cent. The proportion of the population without access to safe water gives an indication of the very low level of infrastructure development in most of these countries. Only in Botswana, Namibia, Zimbabwe and South Africa are a large majority of the population estimated to be within reach of safe water.

Table III Poverty and human development

	Population with income below $1 a day (1985 PPP$ (purchasing power parity dollars)), 1989–1994 (%)	People not expected to survive to age 40 (% of total population 1997)	Population without access to safe water 1990–97 (%)	Underweight children under age 5 1990–97 (%)
Angola	n.a.	38.3	69	26
Botswana	34.7	35.0	10	17
Kenya	50.2	29.8	47	23
Lesotho	50.4	25.1	38	16
Malawi	42.1	47.8	53	30
Mozambique	n.a.	39.8	37	27
Namibia	n.a.	30.0	17	26
S. Africa	23.7	23.4	13	9
Tanzania	16.4	35.5	34	27
Uganda	50.0	47.4	54	26
Zambia	84.6	46.9	62	24
Zimbabwe	41.0	39.8	21	16

Source: UNDP, *Human Development Report 1999*, Table 4, 'Human Poverty in Developing Countries'

Aims of the book

National level data conceal great regional and local variations in poverty and deprivation, as well as inequalities between households and within them. Macro-level analysis can obscure widely differing local processes of change, as well as household level decisions and strategies. It can encourage an approach to understanding rural Africa that seeks typicality, rather than looking for patterns within diversity by exploring comparisons and contrasts, as this book aims to do. My intention is to show the extent of diversity between and within African rural economies and to explain why these diversities are so marked. It also shows where similarities can be found in the problems people face and the responses they develop to deal with them. The intention is to generalise where this is possible and to point out diversities where they are important. The book also aims to inject an understanding of longer-term processes of economic and social change into discussion about Africa's economic 'crisis'. Just as the economic problems besetting African countries did not suddenly appear in the 1970s, so the responses people have made to these problems, the economic, social and cultural resources they have drawn on, have long histories and can be understood only by taking a long-term perspective. Many of the supposedly new processes coming out of rural Africa today – 'multiple livelihoods', changing gender relations – look less novel when viewed historically. Yet there are also discontinuities, new problems and attempts to deal with them in new ways.

States, farmers and migrant labour

The focus on eastern and southern Africa is prompted by some commonalities between these regions. European settlers took over much of the high potential agricultural land and pushed Africans into increasingly crowded reserves. In some of these reserves, many Africans became involved in commodity production. In all of them rural livelihoods depended heavily on labour migration to large-scale farms and plantations and to the towns. State policies later diverged. In the 1950s, the British colonial state in Kenya reversed its policies towards African agriculture, moving away from repression towards the promotion of smallholder farming through opening up farmers' access to high-value cash crops. The post-colonial state maintained this stance. Land redistribution led to a further expansion of smallholder farming, though the really significant increases in production happened in the former reserves as Africans combined commercial farming with labour migrancy. The network of marketing boards which controlled agricultural pricing into the 1980s lowered farmers' incomes considerably (Bates 1981, 1989). In Zimbabwe, the colonial state gave some limited encouragement to African 'master farmers' in the 1950s. The unilateral declaration of independence by the Smith regime in 1965 and the liberation war that followed meant that rural dwellers had to wait until the advent of an independent Zimbabwe in 1980 before the state began a slow, and subsequently much postponed, process of land reform. Smallholder commercial farming remains limited in an agricultural economy which is still dominated by the large-scale commercial farming sector. In South Africa, the dispossession of African farmers went farther than anywhere else in Africa, though it was never complete. During the twentieth century, those Africans who lived and worked in white farming areas were deprived of secure access to land and most were shunted off into reserves. The last wave of this process followed the coming to power of the National Party in 1948, in which several million forced removals took place. Chapter 2 outlines successive attempts by the bantustan state of Bophuthatswana to create a class of African commercial farmers in the later years of the apartheid era. It draws on my research in North West Province to assess the prospects for successful implementation of land reform and rural development.

In those countries which did not experience large-scale settlement by European farmers in the colonial era, rural people were drawn into the cash economy through growing crops for sale and through migrancy to plantations, to towns and, across much of southern Africa, to the South African mines. Migrancy has remained a centrally important strategy.

Changing livelihoods

The repercussions of declining job prospects and living standards in the urban areas since the 1970s have been felt across eastern and southern Africa. Urban real wages in Tanzania fell by 83 per cent between 1974 and 1988. They continued to fall in the 1990s. Many urban dwellers responded by leaving wage employment for self-employment and urban farming (Tripp 1997). In Kenya, real wages for workers in public administration fell by almost a quarter between 1976 and 1989, while real wages in manufacturing fell by 42 per cent between the early 1970s and the late 1980s (Jamal and Weeks 1993). In 1984, it was estimated that two fifths of urban families in Zambia earned less than they needed to pay for their minimal needs (Loxley 1990). The downturn in urban economies has led to a slowing of migration from rural to urban areas and to return migrancy. Population growth rates in urban areas have dropped (Becker *et al.* 1994; Baker and Pedersen 1997; Potts 1995, 1997, 1999; Satterthwaite 1996 - cited in Potts, 1999).

Changes in the South African mining industry have had repercussions throughout southern Africa. Falling gold prices and the restructuring of the industry have led to large-scale job losses. The yearly average total of miners employed on the South African gold mines fell from 477,397 in 1986 to 324,441 in 1992 (Crush 1995: 22). By 1999, in the face of a collapsing gold price, this figure may have fallen to around 280,000 (*Mail and Guardian*, Johannesburg, 'No end to miners' pain', 9 July 1999). Similarly, mine labour employment in the Zambian copperbelt was devastated by the falling copper price from the early 1970s onwards. In 1970, the mining industry accounted for 42.9 per cent of Zambia's GDP. By 1991, it accounted for only 7.3 per cent

(Loxley 1995: 133). The type of labour sought by the South African mining industry has also changed since the 1970s. The major mining houses have stabilised their labour force. This has involved replacing a large pool of relatively low-skilled and low-paid labour on one-off short-term contracts with a smaller group of highly skilled and more highly paid labour on rolling contracts.

> Mine migrants have become a relatively privileged absentee rural elite in the midst of abject poverty. As Basotho miners point out, there is only one thing worse than having a mine job in South Africa and that is not having one.
>
> (Coplan 1993, quoted in Crush 1995: 25)

The impact of these changes is hard to overstate. Many rural households within South Africa and in the wider region have lost their main source of livelihood. They have had to fall back on a medley of activities – small-scale farming, agricultural labouring, petty trading – none of them very remunerative. Their predicament resembles that of households in migrant labour economies farther north whose able-bodied members cannot find adequately paid work in the towns.

If there are similarities between households and regions caught by the decline of remittance economies, there are also many differences, within regions and between them, as well as between households in the same region, in the livelihood repertoires households call on and the long-term strategies they follow.

Many rural households in eastern and southern Africa regularly sell their crops and some grow cash crops, like tea and coffee, exclusively for sale. Farmers bring to these activities a wide range of resources, values and expectations. They are also subject to quite different incentives and constraints from the institutions, state and private, which mediate their involvement in markets in different countries.

Organisation of the book

People living in the African countryside face multiple challenges – economic, political, environmental and social. However, this book will show that there is not a uniform 'African agrarian crisis'. There are marked differences in livelihoods and in levels of poverty between regions, and within them, and many different kinds of local economy. In some regions, there is a relatively flourishing commercial agriculture, with thousands of small farmers making a living from growing crops for sale. In others, rural households which long depended on labour migrancy increasingly put together multiple livelihoods.

The first half of the book is organised around case studies of livelihoods from different types of rural economy. These range from the commercial farming regions in Kenya and Tanzania, discussed in Chapter 1, to the much poorer regions of Kenya, Southern Zimbabwe, Northern Zambia and South Africa discussed in Chapters 2, 3 and 4 where multiple livelihoods predominate. In order to make comparisons meaningful, and manageable, I do not focus explicitly on changing livelihoods in urban economies, other than with respect to their impact on people living in rural areas. This focus does not mean that we should regard urban and rural areas as separate economies, or social fields. They are inextricably linked through migrancy, through markets and through the integrative activities of the state and media. My focus is on the rural dimension of multi-dimensional networks of people, ideas and resources. The impact of changes in the urban dimension of these networks is stressed throughout. It is brought into focus in Chapter 3 and in the second half of the book.

In Chapter 1, I explain how the smallholder farming sector in Kenya expanded so dramatically that it became a model for rural development elsewhere in Africa. I also show that this is not a model that can simply be imported. Kenya's smallholder boom was the outcome of specific circumstances and it happened in a way that was quite different from the intentions of the colonial state. Nevertheless, state policy towards smallholder farming was a crucial element in the story, as the contrast with smallholder tea farming in Tanzania brings out clearly. The South African case in Chapter 2 is based on research I have carried out in Central District, North West Province. It looks at successive attempts by the state to create a class of black commercial farmers in one of the 'reserves', and the emergence of commercial farming independently of these efforts. It considers the lessons of this history for rural development in South Africa today. Chapter 3 focuses on the effects of sharp falls in the demand for migrant labour and on the diversified livelihoods people construct in response. The case studies show how such livelihoods are actively constructed through ingenuity, negotiation and, often, conflict.

Multiple livelihoods depend on the resources people have to hand or can get access to – money, skills, land, livestock, knowledge about an opportunity for trading, for example. The ways they decide to use these resources – whether to buy another cow, invest in a trading venture or spend money on a child's school fees, perhaps – depend on their values and beliefs, their estimation of how risky these different activities are likely to be, or whether they are having to think short or long term. Decisions about what to do will also depend critically on who in the household is making the decision and what the implications of the decision are likely to be for other household members. This makes gender relations a crucial determinant of livelihood decisions. Chapters 4 and 8 focus on these issues. Chapter 4 looks at the links between gender and livelihoods in different kinds of rural economy and in different kinds of household. Chapter 8 shows how gender and livelihoods have changed over the long term in one community in Western Kenya, Koguta. Men's unwillingness to trust their wives to oversee investments in farming contributed to the long-term decline of the economy of Western Kenya. The research for much of the case material which follows was carried out in the late 1980s and early 1990s, when HIV infection rates were considerably lower than now. The spread of HIV among the economically active population in the countries of Eastern and Southern Africa can be expected greatly to exacerbate the problems of constructing livelihoods that are recounted in the chapters which follow.

The case study from Koguta, which occupies the second half of the book, stretches from the 1920s, when the oldest informants were teenagers, to the late 1980s. Koguta in the late 1980s was a very different place from what it was in the 1920s. Farming had long since declined to the point where almost nobody could hope to maintain themselves from this alone. Most people grew enough crops to feed themselves for only a few months each year. Successive generations had seen the rise of large-scale labour migrancy in the 1920s and 1930s, the new employment opportunities opened up by the Africanisation of the state in the 1960s and the effects of urban unemployment and falling urban living standards on migrants in the 1970s and 1980s. Yet there were also continuities. Rural households had been finding ways to combine farming and labour migrancy, and trying to deal with the problems about authority and responsibility it confronted them with, for three generations. Much of the burden of farm and domestic work had always fallen on women. It was they who felt most keenly the burden of adjusting to reduced circumstances – the shrinking plots of land, and drying up of remittances from the towns. From today's perspective, the decades in which some (never a majority) of migrants could afford to send home the cash needed for food and clothing are an anomaly.

1

Changing livelihoods in Eastern and Southern Africa

1 Making a living
Commercial farming

All agricultural areas in Southern and Eastern Africa have been touched by commercialisation. Farmers often sell some of their crops, buy agricultural inputs or even grow crops specifically for sale. However, there are some regions where farmers are intensively involved in agricultural markets. In the highlands of Central and Western Kenya, on the slopes of the Rift Valley and in Tanzania's North Eastern highlands, farmers grow export crops like coffee, tea and tobacco. These crops need a cool, wet climate. They demand a great deal of labour for weeding and harvesting, but they also bring a relatively high return. In KwaZulu-Natal and Mpumalanga in South Africa, and in lower-lying parts of Kisumu District in Western Kenya, farmers produce sugar cane, often on a contract basis. Coffee, tea, tobacco and sugar cane are all crops that must be sold for farmers to reap a return. Some post-colonial states have also encouraged dryland production of food crops, particularly maize, on land redistributed from white settlers in Kenya and Zimbabwe, in the white farming sector of South Africa and in the former migrant-labour reserves of Northern Zambia.

In this chapter and the next, we see how differences between African rural economies have been shaped by the actions of states, by the operation of markets and by strategies followed at the local level. States have sometimes attempted to stimulate commercial farming by exhortation and force, by trying to create a class of commercial smallholders from above, or by improving farmers' access to markets and inputs. At other times farmers have themselves responded to new market opportunities and become involved in commercial agriculture (Williams 1994). Local strategies of accumulation have influenced rural households' responses to market opportunities and the use they have made of agricultural surpluses. Conflict and bargaining over household divisions of labour have affected agricultural productivity as well as household members' access to food and money.

Kenya: colonialism, resettlement and the smallholder boom

Kenyan experiences with commercial farming illustrate all of these processes as well as bringing out some wider issues about African states and farmers. The Kenyan case shows that the conditions under which smallholder farming could

Figure 1.1 Kenya

expand were particular to one time and place. It does not provide a model to be applied mechanically to other countries. Until the 1980s, Kenya was widely seen as the success story for smallholder farming in Africa. However, the reasons for the spread of commercial farming and the policy implications of this experience were not always understood (Francis and Williams 1993). Kenya's image has since been tarnished by political repression and ethnic conflict. But this image of successful agricultural development was always rather misleading, because a very large number of smallholders were passed over by the expansion of high-value commodities.[1] The smallholder boom took place in areas that were suited, in terms of altitude and rainfall, to the cash crops introduced. Other smallholder areas lost out, as few resources went to raising productivity in the lower-value staples, such as maize and pulses, which could be grown in these areas. Central Kenya was the most favoured beneficiary of the smallholder boom. One reason for this was the fact that it is the home area of the ethnic group that dominated the Kenyan state until the 1980s. It is also the region lying closest to the country's main labour and product markets in Nairobi.

Many smallholder households expanded commodity production using earnings from migrant labour (Kitching 1980; Stichter 1982).

Kenya's smallholder boom began in the 1950s. The colonial state did not intend it to happen. In a classic example of late colonial social engineering, the state had hoped to create a rural world of yeomen farmers and landless labourers that would provide it with a bedrock of political support. This blueprint was subverted by smallholder farmers, who took advantage of the relaxation of growing restrictions to move into commercial production. From 1954 to 1963, the annual rate of growth of marketed output from small farms was 7.3 per cent, and from 1964 to 1970 it was 12.6 per cent (Heyer and Waweru 1976).

Until the 1950s, the priority of the colonial state in Kenya had been to protect European farmers from competition from African producers and to secure labour for the European farming sector. The state restricted African cultivation of the most valuable export crops. These restrictions did not succeed in preventing Africans in the reserves from expanding their production of commodities, but it did exclude them from the more lucrative ones. Instead, they expanded their production of maize, cotton and wattle (Cowen 1975). Labour migrancy was also a critically important component of livelihoods in colonial highland Kenya (Kitching 1980).

During the Depression and again in the Second World War, the state, with a pressing need to earn foreign exchange, paid more attention to promoting commodity production by African farmers. However, this goal was constrained by an official ideology of conservationism that spread across British colonial Africa and South Africa in the 1930s and 1940s (Beinart 1984; Leach and Mearns 1996; Mackenzie 1998). The state imposed soil-conservation regulations and restricted commercial maize growing. These policies also legitimated a fear and dislike of accumulation by Africans in the reserves and the wish to secure an adequate supply of labour for European farms. These restrictions and regulations, together with the expulsion of labour tenants from European farms, fuelled the explosive growth of African nationalism and the Mau Mau rebellion in the early 1950s (Throup 1987; Kanogo 1987; Furedi 1989; Berman and Lonsdale 1992).

The colonial state first responded to Mau Mau with a volte-face in its policies towards the African reserves. From paternalistic and authoritarian attempts to slow down economic change, the state moved rapidly to embrace it, hoping to pull apart the Mau Mau alliance of smallholders and the landless by creating a class of accumulating 'yeoman farmers'. This policy involved an attempt to expand African commodity production while keeping a lid on accumulation and maintaining a labour supply for European farming. The twin pillars of the Swynnerton Plan, published in 1954, were the institution of freehold land tenure and the selective loosening of restrictions on African cultivation of high-value commodities, such as coffee, tea, pyrethrum and dairy products. It was also planned to provide credit and a greatly enlarged extension service to 'progressive' (master) farmers, who were considered the keys to successful smallholder farming. The introduction of freehold tenure by a massive and costly programme of land consolidation and registration was supposed to give farmers an incentive to reduce soil erosion and invest in their farms. By promoting a land market, freehold would facilitate accumulation, while rules against subdivision would prevent renewed fragmentation of holdings.

By the late 1950s, however, the maintenance of a heavily subsidised European farming sector was not considered vital either by the British state or by the political representatives of international capital in Kenya. Both were more concerned to negotiate a peaceful transfer of political power and promote continuity in the basic structure of the economy. They were able to come to terms with nationalist politicians in negotiations for the Independence settlement (Wasserman 1976).

The most important elements of the Independence settlement were the transfer intact of a large proportion of the European mixed farm land to a group of African accumulators and the division of the remainder among smallholders in settlement schemes, as a means of reducing conflict over the distribution of land. At this time, the orthodoxy among governments and international agencies was still that these were more efficient than smallholdings. The chief concern of officials dealing with land transfers was, therefore, the maintenance intact of the bulk of the large farm sector, which was habitually described as the 'backbone' of the economy (Wasserman 1976; Leo 1984).

In 1960, a plan leading to Independence was announced. In order to prevent the run-down of European farms in anticipation of political change, the colonial state announced plans to purchase land for settlement schemes and to assist the buyers of large farms. Land would be pegged at 1959 prices. Loans from a Land Bank would fund the orderly purchase of European farms intact by Africans.

In order to satisfy the demand of Africans for redistribution of land, some farms would be purchased by the state and subdivided into smaller holdings. Two types of scheme were planned. The first would consist of 'yeoman' farms of around twenty hectares, with a target net income of £250, integrated into the large farm sector. 'Peasant' schemes, with farms of around six hectares and a target income of £100 would be established on land contiguous with the African reserves. The state aimed to settle 1800 'yeomen' and 6000 'peasants' by September 1963. Together, the new African owners of the large farms and the 'yeomen' of the settlement schemes would maintain output and stability in the rural areas (Leo 1984). These distinctions were reproduced in later high- and low-density settlement schemes.

Rights to land transferred from Europeans to Africans in the former 'White Highlands' were not based on the historical claims of communities to have occupied the land, cultivated it or used it for grazing in the past. It was sold as private property to people whose purchases were usually financed by government loans. These were rarely repaid in full. The land reforms, like the land registration programme, established property rights as the basic principle of land tenure in independent Kenya (Leys 1975: 66–9).

The 'yeoman' scheme met with a poor response from its target population. Africans were keen to buy up intact the large European farms, either as individuals or as shareholders in land companies and co-operatives. But the kinds of people who had the capital necessary to purchase land subdivided into 'yeoman' plots probably saw more profitable opportunities elsewhere in the economy. The 'peasant' schemes, however, were wholly inadequate in the face of the massive demand for land in the Central Highlands. In response to this pressure, the settlement schemes were extended considerably over the course of the 1960s and 1970s, some 'schemes' merely ratifying existing take overs by 'squatters'. By the end of 1982, the Department of Settlement estimated that 64,000 families had settled on 670,000 hectares. The proportion of land in the hands of smallholders was thus much larger than initially proposed. This happened partly because many of the large farms bought by land companies and co-operatives were quickly subdivided among their shareholders. But the policy makers administering the schemes also realised that many small farmers were drawn from the urban middle class, using farming as a support for their primary pursuits (Leo 1984).

One of the most striking aspects of the economic development of the former European areas has been the difference in productivity between different types of farms and farmers. By the mid 1960s, it was clear that many of the large farms taken over intact were badly managed and their owners seriously in arrears over repayments to the Land Bank of the money they had borrowed to buy the farms. This was in spite of the fact that the bulk of agricultural research, extension and credit was directed at them (Hinga and Heyer 1976). In the 1970s, productivity was much higher on smallholdings than on large farms and, within the large farm sector, it was generally higher on the smaller farms. An important reason for this was the tendency for smaller farms to have a larger proportion of land under crops. Smaller farms also employed more workers and sold more crops per acre than larger farms (Livingstone 1986). In the mid 1970s, the state had to confront these problems. It introduced schemes to rehabilitate selected large farms, while a number of large farms were eventually allowed to subdivide into smallholdings. Some under-used farms have also been taken over by squatters. But subdivision of the rest of the large farms will not be politically possible The owners are too powerful (Heyer 1981). The expansion in the proportion of marketed output coming from small farms has not risen above the 1967 high of 50 per cent.

An important part of the explanation of productivity differences between small and large farms lies in the strategies followed by different types of farm owners. What seems to have been happening is that many of the large farms transferred intact, together with a large proportion of farms in the low-density (larger plot) settlement schemes, were purchased by members of an emergent class of African accumulators whose interest lay in the urban economy. In a study of high and low-density settlement schemes in Nyandarua in the early 1970s, Leo found that many settlers in the low-density scheme were absentees, or were involved in other, full-time activities off the farm (Leo 1984: 158–68).

Many people with resources to buy into the low-density schemes regarded their farms as one investment among others, rather than as their chief source of livelihood. They were also launching themselves into many other activities in the post-Independence economy, particularly white-collar employment and business. In consequence, a significant number were neglecting their farms. 'Telephone farmers' were often reluctant to give the managers of their farms sufficient authority to allow them to farm properly.

In the high-density scheme, the few settlers with economic interests off the farm were the handful of people who had held white-collar jobs before joining the scheme. The rest were wholly involved in farming. Leo quotes an agricultural official in another scheme who thought that the best farmers were those in the low-density scheme who concentrated on agriculture, while the next best were the more enterprising high-density farmers. Among the worst were the low-density owners (Leo 1984: 168). What is not clear from Leo's analysis is the extent to which the more enterprising high-density farmers were those with access to off-farm resources. This is the case in the former African reserves, where off-farm income, particularly from employment, is the chief source of finance for agricultural investment. Nor does Leo indicate how productive farms on the high-density scheme actually were. Settlement schemes in general remained relatively undeveloped, with large amounts of land under-used (Hinga and Heyer 1976).

There was, indeed, a dramatic increase in smallholder commodity production which underlay the expansion of the Kenyan economy in the 1960s and 1970s. This was partly due to the extension of the small farm areas in the 1960s, which was a result of the subdivision of European farms in the years following Independence. However, this expansion was due at least as much to very rapid rates of growth of marketed output in the small farm areas which, by 1967, accounted for 50 per cent of marketed output (Heyer 1981: 106).

In summary, the transfer intact of many large farms and the resettlement programmes proved costly and not particularly successful. Large-scale agriculture ran into serious difficulties in the 1960s, while the most important factor behind the spectacular increases in smallholder production was not the extension of smallholder farming through the settlement schemes, but the expansion of production in the former reserves. So the story is one of unintended outcomes – the subversion of blueprints. The Independence settlement of the land question aimed to preserve most large farms intact while defusing conflicts over land. The large farms transferred intact and the low-density schemes were expected to provide the backbone of the country's agricultural production. In the event, the backbone of the Kenyan 'success story' was smallholder farming. Many of the large farms and the low-density schemes languished, under-used, or were later subdivided. In the former reserves, the Swynnerton strategy for a polarised countryside of yeoman farmers and labourers was subverted through an expansion of commodity production far greater than that envisaged by the social engineers of the colonial state.

The Swynnerton Plan had aimed to make high-value cash crops available only to a minority of 'master farmers'. Far more smallholders adopted export crops (especially tea and coffee) than projected. State agencies and transnational companies directed the expansion of intensive smallholder production of tea, coffee, pyrethrum, sugar and dairy production through their control of marketing and restricted the growth of private trade in agricultural produce. This growth in production was accompanied not by polarisation, but by an apparent entrenchment of the middle peasantry in the former Kikuyu reserves and its expansion into the former European farming areas (Heyer and Waweru 1976; Heyer 1981; Cowen 1981a, 1981b).

The boom in smallholder farming occurred for a number of reasons. A critically important reason was the increasing importance of international capital, and the declining influence of European settlers, in the political economy of Kenya. Associated with these changes were shifting patterns of class formation and class alliances. These underpinned settlement of the land question at Independence and also facilitated large inflows of investment in transport, processing and marketing. Second, many smallholders were able to expand commodity production by combining commercial farming with wage labour in multiple livelihood strategies. These factors will be examined in turn. Finally, the smallholder boom happened at a time when external and internal commodity markets were buoyant (Cowen 1981a; Heyer 1981).

The state and international capital

State agencies and transnational companies invested in infrastructure and controlled marketing (Heyer 1981; Cowen 1981a). Central Province was the most favoured beneficiary of this investment. The Province is the home area of the Kikuyu, the ethnic group which dominated the Kenyan state until the late 1970s. It is also the region lying closest to the country's main labour and product markets in Nairobi. But there has also been a considerable growth of commercial production by small farmers in some parts of Western Kenya and the Rift Valley. Much of this commercial production has taken the form of contract farming.

Contract farming is now a common relationship between farmers and agribusiness in developing countries, particularly in Sub-Saharan Africa and Latin America. It is also widespread in the farm sector in industrialised countries. Landholders contract with crop purchasers to sell crops under specified conditions (such as prices, quantities or qualities). Typically, crop purchasers provide seeds, fertilisers and other inputs (such as tractor ploughing) and stipulate crop production methods and timings. The contracting body may be an agribusiness firm, or the state, or it may take the form of a management agreement or joint venture between the state and a private firm.

It is sometimes argued that contract farming is most likely to be found in markets where the crop needs rapid and capital-intensive processing or transportation (tobacco, sugar cane, horticultural crops), or where the crop is produced under highly labour-intensive methods which cannot easily be mechanised (Binswanger and Rosenzweig 1986, cited in Watts, 1994). However, the same crops are also produced under a variety of different production relations, including plantations. Watts argues that the introduction of contractual relationships in a region is never a purely technical matter. The particular histories of different regions and enterprises have shaped crop production and marketing relationships (Watts 1994).

In Africa, the greatest numbers of smallholder contract farmers are found in Kenya, where nearly 350,000 were registered in various schemes in 1991 (Jaffee 1994). A large proportion of these (230,000 in the 1980s) grow tea under contract with the Kenya Tea Development Authority (KTDA), one of the most enduring and successful contract farming schemes in Africa. Other farmers grow sugar cane and tobacco under contract.

The proliferation of contracting in the agricultural sector has been explained in terms of a global restructuring of food production and marketing (Watts 1994). Open-market exchanges between producers, input suppliers and purchasers of farm produce (together with direct control of labour through plantation production) are replaced by contractual relationships that link and regulate production and exchange. There has also been a shift in global agricultural trade towards high-value commodities. This reflects changing patterns of demand in industrialised countries, as well as the pressures Structural Adjustment Programmes bring to bear on developing country governments to promote export crops.

However, the shift towards contracting is not new. Nor is it inevitable. Jaffee traces a history of smallholder contract farming in Kenya's horticultural sector that stretches back to the Second World War. A project producing dried vegetables was initiated then by the Agriculture Department, and there have been more than two dozen contract schemes since (Jaffee 1994). The horticultural sector has grown since the 1970s and it has become an important source of foreign exchange and employment. Contract farming nevertheless seems to have been a temporary stage in the development of much of this sector. Some contract schemes (notably for French beans) have been successful, but many failed, or were replaced by other production arrangements. Contract schemes with smallholders to produce canned pineapple, passion fruit and cut flowers were eventually replaced by production on large farms or on centralised estates. Some contracting firms collapsed. Contract farming in horticulture was actually less significant in the 1990s than it had been in earlier decades.

Contract farming is an often unstable relationship, because it sets up a series of contentious issues for the contractor and for members of the farm household. From the point of view of the contractor, the aim of the contract is to strike a balance between shaping the labour process and final product as much as possible and keeping down the costs of monitoring. This principal-agent problem leads to contracts that specify production methods and product quality in great detail. Contract farming is intended to increase the power of the input supplier/commodity purchaser over the farmer and over the production process, in comparison with the open market, and it usually does so. But problems of enforcement remain. Farmers may not comply with required production techniques; or they may sell their output to other buyers. This happened in a number of the Kenyan horticultural schemes. In one carrot-growing scheme in the 1970s, the contracting company was reduced to employing police officers to inspect carrots leaving the area, in order to check whether they were of the same type being grown under contract (Jaffee 1994: 117). Output may be below the required quality. This was a major problem in pineapple-growing schemes. From the point of view of the head of the farm household, the chief concern is to ensure a reasonable return to the labour and land dedicated to the crop. The cost of inputs, specified and supplied by the contractor, may be deducted from the price farmers receive, making it another focal point for conflict.

Selling to competitors may help farmers avoid paying the full cost of inputs, though this may threaten the renewal of the contract, or the scheme itself. One reason why the KTDA has not experienced this particular problem is that its monopoly over tea-processing facilities means that farmers have no other outlet for their tea. On the other hand, they may have other calls on their time. The contract may be part of a wider income-earning strategy, with household members growing other crops, or engaging in off-farm activities. The amount of labour time household members devote to the crop may therefore be much less than the contractor expects (Jackson and Cheater 1994).

There is also potential for conflicting incentives within households. The usually heavy labour-demands of contract farming raise issues over access to labour and crop income within farm households. Contracts are usually drawn up with household heads on the premise that they have control over the labour supply of the rest of the household. This control is rarely absolute, tempered by the need to bargain, or at least to demonstrate to other household members that they will derive some benefit from the crop income. The consequences of all this for farmers contracting with the KTDA are explored in Chapter 4. Advocates of contract farming argue that it provides farmers with access to credit and markets and transfers technology (Morrissy 1974; Williams and Karen 1985; Peperzak 1985, cited in Watts, 1994). Critics argue that it reflects a significant restructuring of the labour process and modes of labour control within agriculture (Cowen 1981a; Roodt 1985; Watts 1994; Little 1994; Clapp 1994). They also point out that the risks of production are off-loaded on to the farmer (Clapp 1994).

The growth in smallholder commercial farming in Kenya was largely confined to highland areas, the Rift Valley (Kericho District) and parts of the West (Kisii District, South Nyanza). Farmers in the favoured smallholder regions of Kenya responded rapidly to the new market opportunities brought about by the abolition of growing restrictions, while market conditions also favoured a positive response. The buoyancy of international demand for agricultural commodities and in the urban labour market in the years immediately following Independence strengthened the viability of long-existent strategies of straddling farming and wage labour. Many men in Central Province, in particular, were well-placed to work in Nairobi while supervising their wives' labour on the farm.

Multiple livelihoods

This pattern of combining farming with involvement in the labour market, 'straddling', can be found right across the agricultural regions of Kenya (cf. Orvis 1985; Seppälä 1995). Many rural households have used remittances from migrant household members, or other forms of off-farm income, to finance agricultural investment (Kitching 1980; Tiffen *et al.* 1994). It is a common pattern across Sub-Saharan Africa and the phenomena of straddling and multiple livelihoods will be discussed in Chapter 3 (Low 1986). Urban dwellers still try to maintain a rural base to return to when they are ill, unemployed or retired (Potts 1995). 'Straddling' allows people to spread risks and is a widespread response to the uncertainties of constructing a livelihood (Berry 1993: 62–3).

The consequences of straddling for agriculture are double-edged. Straddling lies behind the dynamism of agricultural growth in commercialised farming areas, providing rural households with capital to invest in farming and allowing them to avoid relying on formal credit (Sender and Smith 1986; Heyer 1981). But it also dissipates people's energies, by preventing them from concentrating their attention, time and resources on any one activity. Kenya's telephone farmers illustrate a common predicament for Africa's commercial farmers.

Multiple livelihoods constructed around migrancy can generate farm investment only when the household members earning an income away from the farm can deal with the problem of labour supervision. This issue often has a gender dimension. A male migrant may send money home to pay for food, clothing and school fees, but he will be prepared to sink money into seeds, fertilisers and ploughing only if he is confident that the investment will be properly managed. In many regions, migrant men are deeply suspicious about their wives' commitment to making long-term investments, at the expense of more immediate consumption needs. There may well be a clash of interests between the wife's need for money to buy food and household goods and the husband's wish to build up the farming operation in preparation for his retirement (see Ferguson 1992 and the discussion on pp. 167–8). The male migrant with money to invest faces a principal-agent problem that he usually can solve only by investing in off-farm activities or by trying to supervise his wife's work on the farm. So migrancy can more easily be combined with commercial agriculture in regions lying close to major labour markets, as is the pattern in Kenya.

Commercial farming usually depends on farmers having access to other sources of livelihood, to provide investment capital, to cover seasonal shortfalls in income and to spread risks. Full-time farming also requires a great deal of co-operation between household members. They may not be persuaded that they will all benefit from dedicating their labour wholeheartedly to commercial production, as we shall see in Chapter 4. The feasibility of full-time smallholder farming has been a powerful myth in the history of African rural development policy. The Swynnerton Plan's blueprint for the rural social structure was based on somewhat arbitrary calculations about the amount of income needed to make full-time farming an attractive alternative to wage labouring. At the same time, the Tomlinson Commission in South Africa engaged in similar calculations in its plans for restructuring black agriculture in the reserves (Tomlinson 1955). Kenya's resettlement schemes echoed Swynnerton in their plans for promoting full-time farming. There have been echoes of this myth in post-apartheid debates about, and policy for, land reform in South Africa.

Chapter 3 will show how multiple livelihoods are also becoming common in less commercialised rural areas. There, however, they are a response to the inability of any one source of income to provide households with an adequate livelihood. People may conjure a living from farming, remittances, small-scale trading, artisan work, paid farm labouring and beer brewing. They may switch between these activities as seasons and opportunities permit, with a distinctly improvised air. People respond to opportunities provided by the resources they have to hand, by market demand, by the vagaries of cross-border trade or by social networks, as Chapter 3 will show. The flexibility and improvisation should be celebrated, but they do raise questions about sustainability.

As we shall see in Chapter 4, multiple livelihoods set up difficult issues for intra-household relations. In some contexts, they give rise to interdependencies, as when men in a commercial farming region need access to women's farm labour, while women need access to their husbands' wages and to the crop income. These interdependencies may provide husbands and wives with incentives to co-operate, or they may simply define the points of conflict. Where multiple livelihoods are an improvised response to poverty, they are far less likely to generate interdependencies and the pressures for households to fragment may be very great.

Rural economies are not just about farming. Farm households need multiple sources of livelihood. To flourish, they also need a buoyant and supportive non-farm rural economy to provide them with inputs, services, local employment and local demand. The fact that such a sector cannot spring up overnight is one of the obstacles facing attempts to create smallholder farming from above.[2]

King's study of the informal (jua kali) sector in Githiga village, Kiambu District, shows what a flourishing rural non-farm sector can look like (King 1996).[3] King first studied the rural jua kali sector in Githiga in 1972. In 1992 he found around 150 jua kali enterprises in Githiga, carrying on thirty-eight different trades. They included metalworkers,[4] motor mechanics, cycle repairers, electricians and someone selling insurance, as well as the more typical bars, retailers, tailors and shoemakers.

The metalworkers were producing household goods, building components, spare parts for road vehicles and agricultural machinery and tools. When King first visited Githiga in the 1970s, metalworking was a very specialised skill known to only a handful of people. By the early 1990s, a technical community, with a pool of knowledge and shared information about techniques, raw materials, sources and markets, seemed to have emerged. Githiga metalworkers followed several different paths into the trade. Some learned their skills from working for Mutang'ang'i, a Githiga man who had himself acquired skills from his father and from a six year informal apprenticeship with an Italian foreman on a European farm in the 1940s. Others used a job in the formal sector as a base for part-time self-employment and gradually moved into full-time metal-working. King found the metalworkers to be more technologically advanced than in the 1970s (one reason being that they now had access to electricity). They showed confidence in acquiring and using new technologies. The jua kali sector was also much more diverse than in the 1970s.

In 1974, Mutang'ang'i used hand-operated machinery to make bicycle carriers and foreguards and aluminium serving spoons. He also made fence-post nails from scrap wire and a fodder-dicing machine that used a fly wheel. He was very anxious to get power tools. In 1977, using credit from the Industrial and Commercial Development Corporation, he bought electrical grinders, drilling machines and power presses. He used these to make further machines, including electrically powered machine tools and an electrically-powered fodder cutter. This 'zero grazer' was used for intensive dairy production by farmers who, because of land shortage, kept their grade cattle in paddocks. It provides a good example of positive links between agricultural intensification and the beginnings of industrial specialisation. By 1984, Mutang'ang'i was exporting shovels to Tanzania and needed an accountant to keep track of his flourishing finances.

(King 1996: 150)

A post office, several clinics and good local transport all allowed people in Githiga to find locally what they would have needed to look for in Kiambu Town or Nairobi twenty years earlier. King argues that all this activity suggests that the local agricultural surplus has risen substantially.

Lessons from Kenya

Kenya is far from being an unmitigated success story, as the case study from Koguta will make clear. The smallholder boom was confined to areas suitable for high-value crops, while there have been productivity problems associated with gender conflict in the commercial smallholder areas (see Chapter 4). Nevertheless, the World Bank has cited Kenya's land reform programme as a model for market-based land reform in other countries (Binswanger and Deininger 1993; Williams 1996). One implication of the Kenyan experience is that it is difficult to create commercial smallholder farming from above. Another is that commercial smallholders may emerge from below if they can get access to markets, if there is a supportive institutional environment and, crucially, if circumstances make it possible for them to combine farming with other sources of livelihood. Case studies from North East Tanzania, in this chapter and North West Province, South Africa, in the following chapter, underline this conclusion.

A comparison between the experiences of smallholder farmers producing export crops in Kenya and Tanzania reveals striking differences in the organisation of the industry, in state policy, in farmers' strategies, and, for these reasons, in agricultural productivity.

Constrained commercialisation: highland Tanzania

Agricultural policy in colonial Kenya was dominated by the imperatives of settler farming. Administered by the British under a League of Nations Mandate from 1920 to 1946 and as a trusteeship territory of the UN after the Second World War, Tanzania did not experience widespread land alienation and the population of European settlers was small. Apart from some plantations,

Figure 1.2 Tanzania

there was little investment by private capital or the colonial state. Some densely populated, well-watered highland areas did undergo commercialisation, as African farmers responded to opportunities to grow high-value crops. In most other regions, agricultural commercialisation was minimal. State intervention in agriculture began on a large scale after the Arusha Declaration of 1967, with the implementation of the *ujamaa* programme of villagisation. The aim of the programme was to foster self-reliance in production while at the same time halting the supposed drift towards agrarian capitalism. The failures of *ujamaa* and villagisation have been well documented elsewhere (Cliffe *et al.* 1975; Ingle 1972; Shivji 1975; von Freyhold 1979; Hyden 1980; Coulson 1982). Sender and Smith show how subsequent state agricultural policy has been a major constraint on the expansion of commercial farming (Sender and Smith 1990).

Lushoto District lies in the West Usambara highlands, in the north east of Tanzania. Like the highland commercial farming regions of Kenya, Lushoto's high rainfall makes tea production possible. As in these other regions, too, population densities are high and rising, leading to problems of land shortage and soil erosion. Male out-migration is high, with male-female ratios of 60–100 in some of the divisions where Sender and Smith conducted their research in the mid 1980s. However, there are also some important differences from Kenya. Sender and Smith argue that tea production has been far below potential and much less than in Kenya and Malawi. Lushoto tea growers are also comparatively inefficient, with yields and quality of their tea output much lower than in Kenya. The accumulation process has been much less dynamic in Tanzania than in some other African countries.

Sender and Smith set out to explain the constraints on accumulation processes through micro-level research on rural households and local institutions and through an examination of the impact of state policy on these processes. Together, these processes have stifled the emergence of capitalist agriculture in the region, something that Sender and Smith consider to be essential for reducing rural poverty. Arguing that social change would be most visible at the extremes of the income distribution, they focused their research on the wealthiest and poorest households.[5]

They found that accumulation by wealthier farmers was rather limited. A large part of the explanation for the poor performance of export crops in Tanzania lies in inappropriate policies and institutional weaknesses. An over-valued exchange rate in the 1970s and 1980s discriminated against export crops, while pricing policies involved rising margins for marketing institutions, lowering returns for producers. Tanzania's attempt at self-reliant industrialisation led to declines in production in critical areas of manufacturing (intermediate goods and incentive goods). Despite subsequent devaluations, there are still severe non-price constraints. The tea sector has been plagued by transportation and processing problems, while delays in payments are common. These problems originate in the structure and operating methods of the Tanzania Tea Authority and other export crop parastatals. The authors point out that reforming prices alone will not remove this problem: there is a need for active policy interventions and investments.[6] But Sender and Smith also show how the slow growth of tea production has also been brought about by individuals' accumulation strategies and by gender conflicts within rural households. When they looked at the small minority of better-off farmers in Lushoto, Sender and Smith found striking differences between accumulation strategies, so much so that these seemed to show two quite different social and cultural orientations.[7]

Farmers following 'pre-capitalist' strategies were not investing in high-value crops or innovation. Following a much older pattern, they practised polygyny and relied on capturing the labour of wives and children in order to accumulate resources (cf. Guy 1990). Capitalist strategies involved buying land, investing in higher value crops, using purchased inputs and innovations and employing wage labour. Farmers following a capitalist strategy marketed a larger proportion of output, invested in off-farm enterprises and in their children's education. The majority of this group had experience of high-status employment, while none of the farmers following a pre-capitalist strategy had done this kind of work. These differences in strategy had economic importance, in that the capitalist farmers were following the 'straddling' strategy already described for Kenya. However, the capitalist farmers were also socially and culturally distinct. The fact that most were not polygynists suggested that they had the ability to resist the pressures to marry again and have more children. A disproportionate number of them had educated mothers, and had married educated women. On the other hand, the group following pre-capitalist strategies demonstrated that these were still feasible.

People engaging in these two kinds of strategy were in competition with one another, in the land market, the labour market (increases in the availability of wage employment undermine control over labour through institutions of marriage and kinship) and politically. Getting access to the resources available through the state was crucial to this competition. State expenditure (on education, extension, health, input subsidies) disproportionately benefited wealthy farmers, particularly capitalist farmers. The same was true of opportunities for smuggling brought about by the growth of parallel markets. The outcome of the struggle between these two subcultures would be decisive for the future path of economic change in the region, but it was difficult to tell which one would be more likely to flourish. Nevertheless, it did seem to be hard for farmers following a pre-capitalist strategy to transmit advantage to the next generation, because of the need to raise money for school fees (compare the case of Peter Odiyo in Koguta, pp.134–5).

Sender and Smith found widespread and deep poverty in Lushoto. Most of the households they surveyed had almost no possessions. Yet commercial farmers suffered from a shortage of wage labour. This paradoxical situation is common in other parts of Africa (Sender and Smith 1986). The solution to the puzzle lay in the finding that wage workers came disproportionately from households with no married men in them, or which lacked access to male economic support (such as through remittances). Female-headed households had very limited access to land. The authors speculate that there might be a common process through which households became deeply impoverished. Below a certain landholding size, men might make a judgement that the household could not support their own subsistence needs, desert and sell more of the household's land. Such households would then have to look to wage labour for survival. In male-headed households, women were not free to do wage work. Their husbands, wanting to appropriate women's labour for work on the farm, were not willing for them to look for wage work. So, in both cases, gender was a crucial determinant of poverty.

The study shows the central importance of the availability of rural employment for the poor, particularly for women. The opposition women faced from husbands to their seeking wage employment should not be romanticised as 'resistance to proletarianisation'. Domestic relations in this region were often characterised by violence. However, they may have been more amenable in the long run to economic pressure than was apparent in the mid 1980s. When Jambiya carried out research in Lushoto in the late 1990s, he found that men were actively encouraging women to look for work (Jambiya 1998).

Poverty was a direct effect of there being an insufficiently dynamic rural economy in Lushoto. Sender and Smith believe that more rapid accumulation, and a more rapid increase in wage employment, would offer a much firmer base for improving the living standards of the rural poor than prescriptions from a 'pro-peasant' or 'family farm' perspective. Micro-level responses by the poor are hampered by the constraints on accumulation faced by the better off.

What can be learned from the contrast between these two, very different experiences in Central Kenya and Lushoto District? In Central Kenya, the smallholder boom was made possible through the opening up of markets and the injection of international finance. Farmers used remittances to invest in commercial production. In Lushoto, off-farm income and also employment experience made it possible for a minority of capitalist farmers to accumulate, but this process has been hampered by policies constraining accumulation. However, the lesson from this contrast is not simply that markets are better than states at promoting commercial farming. We have seen that the Kenyan smallholder boom happened under a set of circumstances that was specific in time and space. Farmers in Western Kenya, living under the same policy regime, have not benefited from the boom. The state also played a central role in the smallholder boom in Kenya. Parastatals controlled marketing throughout the period of rapid growth and continue to do so (Bates 1989; Raikes 1993). Sender and Smith argue that what is needed is not a blanket neo-liberal approach, but rather state intervention to promote capitalist accumulation. Because wage income is so important for the rural poor, this approach is likely to be a more effective response to rural poverty than income or asset distribution.

> Poverty alleviation programmes involving redistribution from above cannot realistically be expected to do more than marginally improve the destitute

conditions of life for more than a tiny proportion of the poorest rural inhabitants. Neither the political nor the material basis for effective and sustained redistributive policies exist in Tanzania.

(Sender and Smith 1990: 138)

In other words, you cannot have redistribution with such a small productive base. This position stands on one side of a longstanding debate with neo-populists about the potential small-scale farming holds to reduce rural poverty and underpin a dynamic agricultural sector (Harriss (1982) and Shanin (1988) provide useful edited collections covering this debate). The issues it raises cannot be resolved at an abstract level. Sender and Smith's position rests partly on their pessimism about the ability of the Tanzanian state to distribute welfare equitably. The point has already been made that the smallholder boom in Kenya was the result of a fortuitous combination of circumstances.

Inefficient parastatals reduced the productivity of tea farming, but social relations at the local level and the livelihood strategies these give rise to have also affected productivity. Gender conflicts in poor households lowered the supply of wage labour and competition between farmers following capitalist and pre-capitalist accumulation strategies restricted capitalist farmers' access to state resources.

Local economies in other densely populated highland areas in Tanzania show the importance of local opportunities and the strategies rural households follow in response to them, as the contrast between the Uporoto and Uluguru mountains in Tanzania reveals. The Uporoto Mountains lie in south western Tanzania. Farmers in Uporoto have been involved in both commercialised farming and labour migrancy since at least the 1920s. In the last twenty years, Uporoto has become a major centre for potato production, dominating the trade in Dar es Salaam (Andersson 1996). Many young Uporoto men migrate over a long distance to the commercial farming areas in the north of Tanzania. They move in order to accumulate the cash they need to build up an independent living as farmers: money for bridewealth and for cattle, housing and agricultural inputs. Many migrants have brought back potato tubers from farms in the north and have started growing potatoes for sale. This process has now been happening since the 1920s. The commercialisation of agriculture in Uporoto has therefore been brought about as part of a long-term strategy in which migrancy and potato farming form successive stages in households' developmental cycles. One stage makes the second possible. But farmers in Uporoto have also benefited from being in the right place at the right time. They sell a large proportion of their output in distant Dar es Salaam. This seems surprising at first, because potatoes have a low monetary value in relation to their weight. What makes it possible is that long-distance lorry drivers travelling between Malawi and Dar es Salaam are often empty when they cross the border into Tanzania. Uporoto's proximity to the border makes it a profitable place for drivers to pick up travelling potato-traders and their produce.

Mgeta division, in the Uluguru Mountains, lies about two hundred kilometres inland from Dar es Salaam. Unlike Lushoto, Mgeta has never been a centre for the production of export crops. It shares with Lushoto and Uporoto a high rate of male out-migration. Many farmers grow vegetables for sale, but the area experiences persistent food deficits and almost all households have to buy food for about half of the year (van Donge 1992). Van Donge's study of agrarian change in Mgeta showed that the decline of agriculture since the 1950s paralleled the closer integration of the area into the outside economy. Growing land shortage, and land degradation, reduced food security. In response, many young adults left the area, mainly for urban areas. Superficially, Mgeta looks prosperous, with many farmers growing vegetables for sale in a thriving local market. But this appearance is misleading, because the food staples in the market are coming in from outside. The money to pay for the food comes from selling vegetables, from small-scale business and from relatives in town. The first two of these sources are rather fragile. Vegetables need fertiliser, which needs to be bought. Land suitable for growing vegetables is scarce and difficult to come by. Little land is bought and sold. Vegetable-traders' profits can easily be wiped out by a glut. Accordingly, there is not an obvious group of successful accumulators and, indeed, no reliable accumulation strategies are available. The poor are susceptible to 'poverty traps'. A poor harvest would mean loss of income from farm labouring which would have been used to buy fertiliser for vegetables; poor households may have to sell food crops in order to pay taxes or buy maize. Many such households consist of a single female and some children, heavily dependent on remittances. They are, essentially, 'guarding' the land for relatives living elsewhere. Marriages are often fragile and many rural households are part of confederations of households spread over different locations. This is because links to people with access to cash, who usually live elsewhere, are more attractive than integration into matrilineally linked groups based on common residence and land ownership.

Sabina has two children. After her divorce from her husband, she returned to live with her mother and now looks after her and farms her land. She sells some of the bean crop she grows and does casual work. The household depends on money coming in from outside. Although her mother's two sisters live nearby, they do not have much contact. Instead, Sabina depends on her brother and sisters who live in towns. Her brother, in particular, visits often and is her mainstay.

(van Donge 1992: 86–7)

Van Donge compares Mgeta with Western Ireland. People seem to have lost confidence in agriculture as a way of life. They look outside for their incomes and security, rather than to local kin and neighbours. They measure one another's achievements in terms of the success of children who have left the area. These are processes that are characteristic of the local economies described in Chapter 3.

Population growth and environmental recovery: Machakos, Kenya

The importance of local strategies and multiple livelihoods is also brought out by Tiffen *et al.*'s study of environmental recovery in Machakos District in Kenya. Machakos lies east of the central highlands, in Eastern Province. In contrast with the highland areas favoured by the smallholder boom, Machakos has a semi-arid climate, with frequent droughts. It also has a high proportion of land that is low potential or marginal by Kenyan standards. Nor has there been as much state-led rural development activity as in higher-potential regions (description taken from Tiffen *et al.* 1994). During the colonial period, and since, the District was seen as a problem, a classic case of environmental degradation, with soil erosion being brought about by rapid population growth and poor environmental management. Colonial officials described how the inhabitants were 'drifting to a state of hopeless and miserable poverty and their land to a parching desert of rocks, stones and sand' (Maher 1937, quoted in Tiffen *et al.* 1994: 3).

Observers in the 1960s were also pessimistic, stressing the limited nature of land rehabilitation and the low standards of agricultural practice.[8] These views reflected a dominant, Malthusian-inspired trend in the analysis of environmental change in Africa. Leach and Mearns have identified a set of myths about environmental change in this analysis. These include the prevalence of overgrazing, the 'desertification' of drylands and soil erosion. This conventional wisdom about the apparently widespread nature of environmental degradation has its roots in the scientific ideas that underpinned thinking about ecology in the colonial period. Many of these rested on concepts of ecological equilibrium that have now been questioned. The conventional wisdom has been sustained by inferences made about long-term processes of change from 'snapshot' views of landscapes. Leach and Mearns argue that such interpretations of environmental change should be approached critically. They form part of a 'development narrative' (Roe 1991) which allowed colonial officials to restrict African farming activities on conservationist grounds. This narrative has been carried over into post-colonial environmental management, continuing to privilege the perceptions and priorities of governments and development agencies, rather than those of farmers and pastoralists (Leach and Mearns 1996).

In 1994, Tiffen *et al.* produced a history of environmental management in Machakos that challenged the conventional view of the District as a 'problem area'. In a remarkable recovery from environmental stress, farmers in Machakos have accommodated their farming practices and livelihood strategies to population growth. In the 1940s and 1950s, Machakos needed famine relief in most years. In the 1960s, returns to labour in farming were lower than unskilled wages in urban areas, making labour migration widespread (Heyer 1967). By the early 1980s, income and welfare levels in Machakos had risen to the middle of the range for Kenya and were well above those in the migrant-labour economies of the western region of the country.

Since 1930, there has probably been a roughly threefold increase in the value of output per capita and a tenfold increase in the value of output per hectare. The authors contrast photographs of the same points in the landscape in the 1930s and in the early 1990s. The earlier scenes show bare hills, scarred with erosion gullies; in the later ones, carefully terraced slopes are dotted with home-steads and show many signs of tree planting. Farmers have intensified production by growing higher value crops and by substituting crops for live-stock production; they have undertaken locally managed measures against soil erosion and locally generated non-farm income has increased as a proportion of total income.

Several processes have played a significant role in this case of successful envi-ronmental management. The authors stress the importance of local initiatives. Population growth increased the labour force and fostered the growth of markets. Both of these processes stimulated agricultural intensification, together with encouraging investments in sustainable technologies and management. Local institutions – self-help groups, schools, shops, churches and so on – helped to convey and process knowledge and capital and made it easier for people to react to the changing economic and political situation. The impact of state action was more indirect. Machakos has never been a major focus for state funding, which has tended to concentrate its rural-development efforts in regions with the highest agricultural potential. Some of the most important effects of state activity on rural development have been indirect, through the expansion of schools, the maintenance of peace and some limited funding of roads. These indirect effects have been extremely important, since livelihood opportunities in the local economy are critically affected by access to markets and by the improvement of human capital (Heyer 1996). Non-farm incomes have also been highly important sources of capital for investment in farming.

These factors have all made investment in farming attractive. What has made it possible has been the availability of non-farm incomes. The farmers who hire labour and who intensify their production methods (by increased inputs of fertiliser, soil conservation, manuring, timely labour) have high non-farm incomes. Many receive remittances from absent household members (Meyers 1982, cited in Tiffen *et al.* 1994). Farmers with little land overcame the problem by intensifying production (Rukandema *et al.* 1981, cited in Tiffen *et al.* 1994). Farmers with limited labour could overcome this problem if they could hire labour with money obtained away from the farm (Ockwell *et al.* 1990, cited in Tiffen *et al.* 1994). In a benign pattern of growth, more inten-sive farming systems have generated labour incomes for the poor on farms and jobs supplying inputs, services and consumer goods to more prosperous farmers. While there are fewer backward linkages to suppliers of inputs in this region than would be found in many parts of Asia, there are close links between farm and non-farm sectors through flows of income and capital within families (Tiffen *et al.* 1994: 176).

Conclusion

The case studies from Kenya and Tanzania show that states cannot create commercial smallholder farming from above. Rather, it has emerged in circumstances where local conditions make it possible for rural households to combine farming with other sources of livelihood. States, and agro-capital, can foster this process through policies that ease access to markets for inputs and produce, but they cannot make it happen if commercial farming is not compatible with rural households' priorities. These priorities are often shaped by long-term investment strategies. The case study from Koguta shows how successive generations of rural households gave priority to investing in education over putting resources into agricultural intensification. As Sender and Smith show, these priorities may also stem from conflict within households over access to labour and income. We shall return to these issues in Chapter 4.

The ways in which states can promote smallholder farming, and the kinds of policies which might best support rural people's efforts to make a living, are critically important issues for rural development in post-apartheid South Africa. In the following chapter, I draw on my own research in one of the former 'homelands' to show what these issues look like on the ground.

Notes

1   Nor does it take into account the experiences of Kenya's pastoralist regions.
2   It will certainly be a problem for attempts to promote black smallholder farming in South Africa. The non-farm rural sector was stunted by the stagnation of rural towns in former 'white' areas following the removal of their black populations to the bantustans.
3   Jua kali (lit. 'hot sun') is a Swahili term coined in Nairobi to describe informal-sector metal workers operating in the open air. Now taken on by the Kenyan Government, its meaning has been widened to cover the informal sector more generally.
4   King's main focus was on jua kali metalworkers in Nairobi. He chose Githiga for his rural research because several of the metalworkers he met in Nairobi came from Githiga.
5   In an important discussion of methodological issues, Sender and Smith criticise household survey methodologies which aggregate and look for typicality, or which focus on small farms as production units. They argue that this kind of approach can lead researchers to overlook those who fail to retain access to the means of production. It can also obscure the dynamics of agrarian change, which tends to happen at the extreme ends of the distribution, around accumulation and loss of assets and changes in relations of production. A random sample may therefore create a false picture of homogeneity. Sender and Smith identified a group of capitalist farmers and also used some imaginative methods to identify the poorest households, including following up the households of school drop-outs. They then constructed a possessions score, which they used as an index of material wellbeing (Sender and Smith 1990).

6   Raikes 1993, writing about agricultural markets in Kisii, Western Kenya, makes the important point that advocating markets in the abstract is meaningless. It is vital to understand how real markets work in practice.
7   They admit that these strategies are often hard to distinguish in particular cases, partly because any strategy takes a long time to come to fruition (see the Koguta life histories).
8   '[S]tandards of agriculture have remained very low over most of the district; virtually no improvement has taken place in animal husbandry', McLoughlin (1967), quoted in Tiffen *et al.* 1994: 15.

2 Farming and rural livelihoods in South Africa

A case study

In the 1990s, debates about rural development policy for South Africa revolved around the prospects for the implementation of a significant land reform programme and whether such a programme could promote sustainable rural livelihoods (Murray 1995; Lipton *et al.* 1996a, 1996b; Levin and Weiner 1997; Williams 1997). Large-scale land reform now looks increasingly unlikely, but the question of what kinds of rural restructuring could support and extend the possibilities for people in the rural areas to make a living remains. Extrapolating from people's current livelihoods is difficult and may be misleading. A recent collection of research on rural livelihoods in South Africa revealed great variation in the importance of agriculture in local economies, while the editors disagreed about the potential for enhancing rural livelihoods through promoting smallholder farming (Lipton 1996; de Klerk 1996).

My research shows why this is such a difficult issue. In North West Province, the former bantustan state of Bophuthatswana engaged in a series of disastrous attempts to create a class of commercial smallholder farmers from above. At the same time, independent black commercial farmers engaged in sharecropping maize on hundreds of hectares of land. Many of these built up their operations with earnings from wage labour and used second-hand machinery bought from white farmers. However, it is far from clear whether the fact they exist at all shows the potential now for an expansion of livelihoods built around commercial farming. Most of them are heavily in debt, while recurrent drought makes both arable and livestock farming a risky undertaking.

The appropriate starting point for rural development in this region must be an understanding of the constraints households face as they construct their livelihoods. Most households lack access to land or the means to work it. They depend on multiple sources of livelihood, in which gaining access to a wage income or a pension plays a central role. Problems of risk dominate people's efforts to construct their livelihoods. Promoting sustainable livelihoods involves reducing the riskiness of livelihoods and reducing people's vulnerability to the effects of risk. It must also avoid increasing the risks they face.

Land reform and rural development

Black South Africans have been deliberately excluded from access to land, capital, employment and education. But there are important differences between regions and within them. Many people lack land; others have land but not the means to work it; for many the most pressing need is for employment. People also put together livelihoods in diverse ways. There are rural areas which offer some of their population a resource base for making a livelihood locally. In KwaZulu-Natal, farmers produce sugar-cane commercially, many on the basis of outgrower or contract production with sugar milling companies (Vaughan 1992). In former Bophuthatswana, the Rustenberg platinum mines provide employment. In other regions, people depend mainly on resources coming from urban areas in the form of wages, remittances and pensions or on seasonal labour on white farms. Settlements range from closely settled rural slums to widely dispersed communities of farmers and pastoralists. Many rural populations are also highly stratified, with income and asset distribution skewed by class, gender, ethnicity and date of arrival. This stratification is bound up with inequalities of voice and power. Attempts to tackle rural poverty need to address these diversities.

The 1994 Reconstruction and Development Programme (RDP – refashioned as the Growth, Employment and Redistribution Programme in 1996) committed the new government to the redistribution of around 30 per cent of medium to high-quality white commercial farmland to black smallholders over five years, through a combination of market and non-market mechanisms. The programme for land redistribution was based on proposals put forward by the World Bank in 1993 and incorporated into the RDP (World Bank 1993; ANC 1994). The programme would combine elements of restitution and redistribution. Communities removed from land in white farming areas after 1913 who could demonstrate rights to land were invited to submit claims for restitution to a Land Claims Commission. Other individuals and communities were to be given access to a programme of land redistribution. Participants would receive a combination of grants and loans for land acquisition and for purchase of equipment and house-building materials.

The restitution programme is a judicial process involving investigation of competing claims to land. So far, only a handful of claims have been settled. In 1994, the Department of Land Affairs (DLA) began a Pilot Programme of land redistribution, to run from 1994–97, later extended to 2000. One district per province was chosen as a site for a Pilot project. Land was to be made available to groups of households, which would jointly purchase land, after agreeing a business plan for use of the land with the DLA and local and provincial government. The Pilot Programme has made some progress, but it is behind schedule, with much of its funding still unspent. Land reform has now been extended to non-Pilot areas, but progress has been slow.

The reasons for this slow progress are political and administrative. There is no powerful political constituency putting pressure on the DLA to speed up the reform process. The most important administrative problems stem from weaknesses in the management of the DLA, with problems in management procedures and information systems. Many staff administering the programme have come into government from the NGO (non-governmental organisation) sector and lack managerial experience. DLA has also found it difficult to co-ordinate its activities with other government departments at the provincial and local levels. There are also flaws in the design of the programme, which may reduce its impact on rural poverty. The impact of the programme depends on: (i) whether beneficiaries are the rural poor; and (ii) whether beneficiaries can construct a livelihood on the land. There is tension between these two considerations, because the people most likely to make a success of farming are unlikely to be the poorest. There is growing concern that the projects may become 'pockets of poverty'. The land reform programme has been criticised for its importation of unsuitable models from elsewhere and for its overconfidence both in the capacity of the state to restructure rural society and also in the potential for land reform to generate livelihoods (Murray and Williams 1994; Williams 1996, 1997; McIntosh and Vaughan 1996).

In 1996, the Government introduced a new layer of elected local councils. These now have responsibility for delivering rural development. In North West Province, District Councils have so far mainly concentrated their efforts on road construction and water reticulation. A broad-based planning exercise is now under way. Elected village committees have been charged to identify priorities for rural development. These priorities are to feed into planning through the drawing up of Land Development Objectives.

There is a strong possibility that it may prove difficult for rural development activity by the state to move beyond the provision of better infrastructure. Many councils are administratively weak, their staff are inexperienced and they are under-resourced. The other obvious players in rural development might be expected to be the DLA and the Department of Agriculture. The DLA is meant to hand over land reform projects to the extension service of the Department of Agriculture once the transfer of land is completed. The Department of Agriculture is attempting to turn itself around from an almost exclusive concentration on the more commercialised farming sector to the promotion of smallholder farming, but this process is still at an early stage, at least in North West Province. Moreover, since so many households lack land and the means to work it, rural development also must be about the promotion and support of livelihoods that do not depend directly on smallholder farming.

Madibogo: the regional context

Madibogo lies in the Magisterial District of Ditsobotla, in Central District, about ninety kilometres from Mafikeng.[1] Madibogopan is an outlier from Madibogo (about 5km away). Very large numbers of people have moved to Madibogo and Madibogopan since the 1960s. Breutz's (1955) estimate of the population of Madibogo, based on the 1951 Census, was 1,440. The 1991

Figure 2.1 South Africa and North West Province

Bophuthatswana Census gave a figure of 22,000, a fifteen-fold increase over a forty-year period.[2] Most of this increase was made up of incomers from farms in the Transvaal, Northern Cape and Free State. The large majority of people do not have access to any land beyond a house site. Many people who do have access to land lack the means to work it. Most agricultural land is sharecropped, by both black and white farmers. Farming is mixed (mainly maize, sunflowers, groundnuts and cattle), but the overall potential is low, because of low and unreliable rainfall. There is no good quality income data for the District. In 1988, the Bureau of Market Research carried out an income and expenditure survey which estimated the components of average household incomes in the District to be as follows:

Table 2.1 Components of average household income in Ditsobotla District

	%
Cash from salaries and wages	39.3
Home production and own business	1.4
Income from farming	6.3
Contributions received	32.9
Other	19.1

Source: Taken from Bureau of Market Research 1989: 15

Note: Percentages do not add up to 100 because of rounding

These figures should be treated with great caution. Not only do inequalities in income make any average figure suspect, but the actual position is that most people earn nothing from farming, while it is a major source of income for a few commercial farmers. However, the data do show a heavy reliance on remittances and pensions that fits with the accounts given to me by informants.

There are no reliable employment or unemployment data for the region. Migrants from the region rely heavily on work in the mining sector, which has been badly affected by retrenchments. The main local employment available is seasonal work on white commercial farms in the region between Delareyville and Vryburg.

Bophuthatswana was declared 'independent' of South Africa in 1977, under the Presidency of Lucas Mangope. The new 'state' was a patchwork of land held under different types of tenure (Jeppe 1980). Its core was formed from 'Tribal' land, administered as African reserves under the authority of chiefs. In the Tribal areas in which Madibogo lies, the Chief allocated housing sites, grazing rights and rights to arable land to households. This is still the case today. However, since the 1960s, very large numbers of people have moved into these areas from farms in the Transvaal, Northern Cape and Free State and have been allocated only a housing site. As a result, the large majority of people in these areas do not have any access to arable land. To the Tribal areas were added 'Trust' land acquired from white farmers by the South African Development Trust between the 1930s and the 1970s. Such land was held by the Trust for redistribution to incoming black communities, many of whom had been subjected to forced resettlement (Platzky and Walker 1985). In practice, a substantial proportion of Trust land has been rented out to black and white commercial farmers. Most households in the Trust areas have no access to arable land. Research carried out in 1988 suggested that only 19 per cent of households in Ditsobotla were cultivating crops (Emmett *et al.* 1988). Since many of the 'cultivators' would be households that leased out their land in exchange for a share of the crop, the actual number of cultivators would be much smaller.

Bophuthatswana was reincorporated into South Africa in March 1994, forming the North West Province with former 'white' farming areas from the Transvaal and Northern Cape. The ANC won an overwhelming victory in

North West in the election of April 1994. While the ANC also won solidly in the June 1999 election, Lucas Mangope won a seat in the National Assembly and his United Christian Democratic Party is now the official opposition in the Province, ahead of the New National Party. This comeback for the former bantustan President surprised observers who had been struck by the incompetence and corruption of the former regime. His success reflects discontent with the retrenchments in the public sector that followed reincorporation into South Africa, with the recession in the private sector that followed on from these retrenchments, and with the slow pace of delivery of promised services.

Around half of Ditsobotla is formed from the former Setlagole and Kunana Reserves. This area contains long-settled communities, including the precolonial Tswana towns of Khunwana, Setlagole, Kraaipan, Madibogo and Lotlhakane. The rest of the District is made up of land acquired from white farmers by the South African Native Trust after the 1936 Land Act, which empowered the Trust to acquire land for settlement by Africans. Some of the settlements on the Trust land are formed from communities which were forcibly removed from 'black spot' areas in the Transvaal from the late 1940s onwards. Others were settled more gradually by people displaced from urban areas and white farms. Additional land was acquired from white farmers by the Bophuthatswana government after 'independence' in 1977. Much of this has been leased to individual, large (mostly black) farmers.

White commercial farmers

The district has to be seen as part of a regional economy in which white commercial farming plays a central role. Ditsobotla is almost entirely surrounded by the white farming districts of Vryburg, Delareyville and Lichtenburg. In Lichtenburg District, many farmers combine maize cultivation with stock keeping. To the west, as rainfall becomes lower and more unreliable, stock keeping becomes increasingly predominant. In the Mareetsane/Vryhof area, immediately north of Ditsobotla, the average farm size is around 1200–1500 hectares. The smallest farms are around 3–400 hectares, and anything under 500 hectares would be considered small. These form a small proportion of the total, partly because many smaller farms went under in the financial crises of the 1980s. Farther west, in the drier, stock-keeping zone around Setlagole, farm sizes are larger. One farmer interviewed had 18,000 hectares spread over 200 kilometres, reflecting a process of land consolidation.

Links between the white farming sector and the former homeland operate through the labour market, through the extensive involvement of white farmers' co-operatives and agribusiness in commercial agriculture in former Bophuthatswana and through the transfer of skills and technology at the farm level (Keenan 1984; Roodt 1985; Stacey 1992; Mather and Drummond nd.). The white farmers' co-operative in the region, Noordwes Kooperasie, together with agribusiness concerns, was heavily and profitably involved in 'homeland' agriculture. Noordwes collaborated with the Mangope government in the form of Agricor, the Agricultural Development Corporation of Bophuthatswana. Together they established state-led farming projects in Ditsobotla, with Noordwes providing many of the senior personnel. Individual white and black farmers also have contact with one another. Black and white commercial farmers meet when they come to the local co-operative for inputs; many black commercial farmers have bought second-hand farm equipment from white farmers (Interview, Lawrence Bodibe, Mooifontein, 21 July 1995).

South Africa's white commercial farming sector has long been heavily regulated and subsidised. From the advent of the National Party government in 1948, white farmers were offered subsidised finance and encouraged to invest in capital equipment. The cheapness of capital also encouraged cultivation of grains well beyond the zone of adequate rainfall. The margins of this zone lie in the Western Transvaal. In many areas in South Africa, this process of capitalisation drastically reduced the demand for labour on white farms. Farmers replaced permanent workers with seasonal and casual labour, much of it done by women and children (de Klerk 1984; Marcus 1989).

One farmer interviewed in Mareetsane in 1995 had a mixed farm of 2900 hectares. He employed only twenty-five permanent workers and sixty to eighty seasonal workers. The permanent workers were all male and the seasonal workers were half men and half women. He actually used more labour than he did before the switch to maize growing that happened after the 1950s, but these figures imply that total demand for permanent farm labour locally is low. Nevertheless, seasonal farm labour remains an important source of livelihood for many rural households. In 1988, 505 farms in Delareyville District employed 11,063 permanent workers, an average of twenty-two permanent workers each (Central Settlement Planning Team 1996: 57). Assuming that they would take on about three times this number of seasonal workers for weeding and harvesting, this would mean that over thirty thousand people would find seasonal work in Delareyville District. However grim this option looks, it does offer a reasonably secure food supply. High unemployment makes farm work look like a more attractive option:

> 'It wasn't OK on the farms. You didn't have time to rest. You had to work right around the clock. We were not well taken care of. Even if we were sick, we had to go to work. ... After arriving [in Madibogopan], we depended on seasonal work on the farms. ... When the children were still young they stayed at home. When they got older, my husband and I stopped harvesting and told the children that if they wanted to make a living they could also go for harvesting. Jobs are so scarce and the transport is free. The children agreed because they realised that after harvesting they would also be given sacks of mielies. They would bring these home and then they could stay for some months, living off that. ... If you don't go harvesting, you will be without food.'
> (Interview with Rebecca Gakiekitsi Senatle, Madibogopan, 19 May 1999)

From the mid 1980s onwards, subsidies to the cost of capital were greatly reduced and farmers have faced rising interest rates, cost inflation and drought. With reductions in subsidies to the cost of capital, some farmers cut costs by switching back to stock-keeping, which economises on capital inputs but which also calls for less labour than maize growing. Many white farmers also face the prospect of insolvency.

Black commercial farmers

African farmers have been growing crops commercially in Ditsobotla for well over a century (Comaroff and Comaroff 1997). Many have been able to produce on a substantial scale, sharecropping several hundred hectares on land whose holders cannot themselves afford to buy or hire tractors. Most commercial sharecroppers produce maize using tractors, planters, trailers and threshing machines. In his study of commercial farming in Ditsobotla and neighbouring Molopo Districts, Stacey estimated that in the late 1980s, 66 per cent of land was sharecropped. He showed how a stratum of sharecropping commercial farmers developed from the 1950s. Many black share croppers bought second-hand equipment from white farmers in the surrounding districts, using earnings from migrancy, sale of livestock and farm income (Stacey 1992; Stacey *et al.* 1994).[3]

My interviews with commercial farmers revealed complex processes of accumulation through kinship, the labour market and access to the local state. Some were born into households of substantial farmers. Others built up their farming enterprises with earnings from trading, or from migrant labour. Some commercial farmers first became involved in farming through providing capital in a partnership with more experienced farmers who lacked capital. They later moved on to farming for themselves. Moeng John Skosana worked as a driver in a number of garages in Delareyville and later in a bakery, finally leaving work in 1986:

> 'My interest in farming was stimulated by a friend, Gaudi, in Delareyville. Gaudi told me that you could farm well in Madibogopan. Gaudi had some equipment and I had money so we combined farming together. ... Gaudi made most of the decisions, because I was from Taung, where they didn't farm. We farmed on Chief Phoi's land and would pay him one bag in five. Gaudi didn't have land. He came from the farms. That is where he learned his farming skills. During the week, we hired people to work on the farm and then we ourselves worked on it at the weekend ... [this led to problems with labour supervision, so Moeng left his job]. Gaudi moved to Sione, so then I farmed alone.'
> (Interview with Moeng John Skosana, Madibogopan, 26 May 1999)

That these farmers exist at all is surprising to many people, in view of the obstacles they have faced in terms of access to finance and the competition from subsidised white farming. To some extent they reflect the survival and transformation of wealth in livestock from a highly stratified pre-colonial Tswana society, although very little is known about the extent of continuities and discontinuities in stratification.[4] In his study of land and class in Thaba Nchu, a Tswana enclave in the Orange Free State, Colin Murray shows extensive continuities between the pre-colonial and the bantustan elites (Murray 1992). It could well be that similar continuities can be traced in Ditsobotla. Access to the local state, particularly farmers' co-ops, has also offered a number of farmers access to critical resources, particularly credit. In addition, the importance of earnings from off-farm activities as a source of capital mirrors rural accumulation in other parts of Africa.

Farming in this region is capital-intensive – requiring use of tractors, other equipment and purchased fertilisers and insecticides – to make it pay. Past attempts at institutionalised equipment-sharing have failed (interview with Adrian Masigo, Assistant Director, Ditsobotla District, Department of Agriculture, Mooifontein, 16 March 1999), and aspiring farmers buy their own tractors. There have been severe problems around the provision of agricultural credit. A number of farmers claimed that they had been pressured into taking unfeasibly large amounts of debt in the past by Agribank, the agricultural bank of former Bophuthatswana. Credit also arrived too late in the season to allow timely planting. Now commercial farmers can apply to the Land Bank, for smaller loans than were allowed in the past. It is too soon to tell whether this reform will allow them to pay back the debt so many have fallen into. Because of the risk of drought, they are on a knife-edge. William Boikanyo left his job as a bus driver and started farming in 1994, sharecropping with two landholders on 97 hectares and 54 hectares. By 1998, he was farming for four people.

'I didn't succeed last year, because of the drought. This year [1999], I ploughed for two people, on 90 hectares and 24 hectares. In 1994, I used my pension money to farm. I made a profit of R22,000. In 1998, I got a loan for R7000. I got only R6000 after the harvest, so I didn't have enough money to repay the loan. This was the first time I had got a loan. In other years I had financed my farming from the previous year's earnings. But the money kept going down and diesel and fertiliser were cheaper in the past. This year I got another loan of R10,000 from the Land Bank. I won't get a profit from this year's harvest because the plants were scorched by the sun. I will wait to hear what the Government says about this. If it says nothing, I'll go and look for a job. … I don't regret getting involved in farming, because I alone don't control it. It's God, because of the rain.'
(Interview with William Pogiso Boikanyo, Madibogo, 27 May 1999)

Access to land may also be problematic, since outsiders wanting to sharecrop may find it difficult to get sufficient land or to enforce contracts. Only through getting access to land through the state can farmers hope to farm on a substantial scale.

Agriculture and rural development

Attempts to create commercial smallholder farming

The local economy consists of black and white commercial farmers, people who hold land without working it themselves and a larger group of people who lack arable land altogether. There is no intermediate group of smallholder farmers. Successive regimes have attempted to create such a group. In 1955, the Tomlinson Commission proposed that freehold titles be granted to full-time farmers on 'economic farm units'. On these, 307,000 black farm households in South Africa would be expected to make an annual income of £120. The rest of the population would have to find wage employment. Since this plan would involve displacing 80 per cent of people from the reserves, the Commission agreed to recommend 'half-sized' units, but these would still involve displacing 49 per cent of the rural population (Tomlinson 1955; Davenport 1987; Ashforth 1990; Francis and Williams 1993). The government rejected these plans and continued instead a policy of 'betterment' planning. From the late 1940s, communities on Trust land in Ditsobotla were subject to 'betterment' planning. Land was consolidated and reallocated for agricultural, grazing and residential use, in the hope of creating a class of full-time farmers. Land right holders, mostly men, were allocated land within the agricultural zone.[5] Betterment was intended to protect the soil and raise agricultural productivity. Instead, in many areas, betterment disrupted people's lives, and reduced the land available to them for farming (Yawitch 1981; Beinart 1984, 1989; de Wet 1989; McAllister 1989).

The Ditsobotla Dryland Projects combined state, private capital and white agricultural co-operatives in the financing and management of contract farming systems. Their main achievement was to leave participants without access to the land they had been allocated and in debt (Roodt 1985, 1988; Reimer 1987). The Bophuthatswana state never resolved the conflict between its commitment to maximising agricultural output and its supposed wish to establish a spectrum of farmers, ranging from subsistence producers to large commercial farmers (Roodt 1985). The Projects were established from 1977 onwards on Trust land in the east of the District. This was therefore the second major reallocation of land in thirty years. In the areas incorporated into the Projects, land which was nominally held by individual participants was cultivated by 'progressive' farmers[6] chosen by the Project management. Many of the managerial staff were seconded from white co-operatives and agricultural consultants. From the inception of the Projects, the majority of the land-right holders ceased to farm. Those who did farm were subject to intrusive and often authoritarian management practices. Farmers received inputs from the Projects and delivered their output to Project co-operatives. Costs, extremely high because of the capital-intensity of production, were deducted from sales. In most years, participants received no returns and often found themselves in debt to the Project after the harvest and people had to look elsewhere to find an income. In 1991, Agricor finally agreed that the Projects could not survive and it was decided to return the land to land-right holders. Attempting to create a class of commercial small-holders effectively dispossessed land-right holders. Top-down management practices made those involved in farming little more than farm labourers. What the projects did achieve was to offer guaranteed markets for agricultural inputs and off-load production risks on to farmers and land-right holders (Roodt 1985).

Local people are now suspicious of state institutions and reluctant to commit resources to community development projects. One of the most difficult tasks facing the new regional government is to establish its legitimacy in communities where state action in the name of rural development has meant only disposses-sion and indebtedness. The ANC began organising in these communities so recently that it cannot assume automatic support. Much will depend on the nature and extent of delivery of projects and their impact on rural livelihoods. In March 1996, the RDP as a separate administrative entity was abolished and the Commission's functions were reallocated to line ministries, although 'lead' projects were to continue to be implemented.

Attempts to stimulate large-scale commercial farming

Long-term processes of accumulation in African farming have intersected with the local state's plans to promote a commercial farming class through the farmers' co-ops, through the Ditsobotla Projects and also through the direct allocation of Trust land to commercial farmers. Much of the land incorporated into Bophuthatswana after 1977 was leased out to commercial farmers. There was a contradiction at the centre of this policy. Mangope's political power rested on his ability to distribute patronage in the form of government jobs, trading licences, business contracts, infrastructure development and land (Lawrence and Manson 1994). Land acquired by the state from white farming areas was a piece of patronage for allocating among his supporters in the bantustan elite. Mangope was also concerned to make Bophuthatswana's 'inde-pendence' from the South African economy more tangible. He wanted to be able to make the claim that the 'nation' was self-sufficient in food production and equated 'development' with large-scale mechanised farming. Agricor policy aimed to maximise 'national' food production, through the Ditsobotla Projects, through irrigation schemes and through the support given to large-scale commercial farmers.[7] Its ability to succeed in this aim depended on identifying and supporting productive, committed commercial farmers, on the 'progressive farmer' model that was common during the colonial period in Kenya and Zimbabwe. The contradiction between patronage and productivity can be seen in the operation of the Land Allocation Board, which allocated leases on land acquired from white farmers after 1977. The Board consisted of the Secretary for Agriculture, personnel from the state banks, Agricor and white farmers appointed from local co-operatives. The criteria with which it chose recipients, ostensibly, were the farming skills and financial position of applicants. Some very large grants were made. In Madikwe, Lehurutshe and Marico Districts, allocations were as large as 3000 hectares. Since many farms were leased out with infrastructure intact, the leases were a substantial resource. The urban elite has been keen to acquire land, but not necessarily to farm it. Many politicians and bureaucrats, including Mangope himself, successfully applied to the Land Allocation Board, which was subjected to heavy political pressure. Mangope sometimes overruled the Land Allocation Board's advice and gave the land to his political allies. Some of the recipients were successful commercial farmers who fulfilled the requirements for commitment and farming skills, undertaking to live on the farm and work full-time. Others regarded the land as one resource within a portfolio and took little active interest in farming. They had diversified business interests which kept them in the urban areas, contrary to the ostensible Agricor ideal of full-time commercial farming. Many farmers have become so heavily indebted that they can no longer get finance from Agribank, which has refused to finance farmers with outstanding debts. Their land lies idle, or is being leased by white farmers while the new government deliberates over what to do with it. The DLA is now charged to sell-off state land and some leaseholders have been offered the option of purchase. Their earlier ability to access land through the Bophuthatswana government is about to be turned into a one-off transfer of land. There are long-term 'squatters' on several farms. The DLA has not yet made any plans for accommodating them.

Attempts to create a group of large commercial farmers through the transfer of farms intact are vulnerable to political pressure. Beneficiaries are likely to be those with privileged access to the redistribution process. Such people are also likely to have commercial interests in the urban economy that lessen their commitment to farming. The bantustan state's attempts to create a commercial farming class from above, therefore, foundered on its inability to match the skills and commitment required for commercial farming with the demands of patron-client politics.

In some households, then, people have kept their farming skills over several generations and have learned new skills through contact with white commercial farming. Even under a discriminatory policy regime, in which black farmers found it very difficult to get access to inputs and markets, some found the resources to invest in and expand their operations. While attempts to create commercial farmers from above in former Bophuthatswana have been disastrous, other farmers have themselves used their skills and capital to engage in commercial farming.

Policy implications

The critical questions concern the scope for an expansion of black commercial farming and the scale of farm operation on which such an expansion might be possible. Additionally, what scope is there for promoting rural livelihoods which combine farming with other activities? What kinds of rural restructuring could support and extend the possibilities for rural people to make a living?

The diversity of the rural sector in South Africa makes generalised responses to these questions impossible. This case study allows us to think about the conditions under which black commercial farming has arisen and how these conditions are likely to be affected by changes in policies and market structures. One important starting point for thinking about possibilities for change has to be an understanding of what people have been doing and why. Economic activities in Ditsobotla District and its surrounds can be conceptualised in terms of a number of margins. These include the margin in both white and black farming areas between maize production and livestock keeping; the margin between white farming areas and black farming areas; and, in the former bantustan, between holding land and leasing land to cultivate. In the past, these margins were shaped by state intervention (subsidy regimes, access to inputs and markets), by the operation of markets and by land tenure regimes. If land redistribution does not take place on a large scale, as looks increasingly likely (Williams 1997), movement of these margins is likely to be decided by market forces, but could also be shaped by institutional changes. It is important to think about the likely effects of liberalisation. That the black commercial farmers exist at all, in the face of the discriminatory regimes of the past, might inspire confidence about how they will perform under liberalisation. The deracialisation of the land market since 1991 will allow a small minority eventually to acquire land in the former white farming areas. Securing tenure rights in the former bantustans may encourage more landholders to lease out land. So there may be some scope for the expansion of commercial farming. Constraints to the expansion of black commercial farming are the availability of capital and the relative returns to maize farming compared with other possible activities, particularly urban employment. Returns to maize farming depend on the world price, on pricing and marketing arrangements for the crop which influence the difference between world, national and local prices and on the cost of inputs. All of these variables are influenced by institutional arrangements which are changing in the direction of liberalisation at present. With improved access to markets for final output, returns for black farmers could rise. However, Stacey's evidence about the precarious financial position of many current sharecroppers suggests caution about the sustainability of large-scale commercial farming in this region. It is also important to take account of the financial problems facing local white farmers. The white farming sector's difficulties are partly the result of over-capitalisation encouraged by cheap credit in the past, but they do also suggest that optimism about the prospects for sustainable commercial agriculture may be misplaced.

If the scope for expansion of large-scale farming in the region may be quite limited, in what other ways could policy promote rural livelihoods? One alternative might be an expansion of smallholder farming. Two starting points for thinking about the potential for smallholder farming are, first, the very limited amount of smallholder farming currently to be found in the district and, second, the failures of previous attempts to create a class of smallholder farmers in this region.

Clearly, rural dwellers in the past faced a discriminatory regime. However, the paucity of small-scale farming already to be found in the district should make one cautious about the scope for future expansion.

One important issue concerns returns to scale. Many white farmers locally believe that there is a minimum threshold farm size for maize production (although they disagree over what it is). The financial difficulties recently faced by smaller white maize farmers underline the importance of this question. The white farmers interviewed were strongly of the view that it is not possible to make a reasonable return at present other than by farming with tractors. But returns to scale are also affected by institutional arrangements, including those governing the availability of equipment. They might be overcome by pooling arrangements, but, as has already been mentioned, these have failed in the past. Disentangling agro-economic constraints from the impacts of apartheid institutions also involves looking critically at the agrarian ideologies which justified them. The previous regime of subsidies available to white farmers encouraged them to extend the margin of maize cultivation into the zone where low and uncertain rainfall makes it a risky undertaking.

In the Lehurutshe District of former Bophuthatswana, Drummond and Manson have traced a history of irrigated small-scale farming which was undermined by discriminatory input and marketing structures, by population relocations and by the loss of access to water (Drummond and Manson 1993). Ditsobotla is much drier and it is difficult to see how the land could be used more intensively. Irrigation would be highly capital intensive and would possibly be captured by those with political access.

This discussion raises questions about the probability of future co-operation between agribusiness and the state in the development of commercial farming in North West Province.[8] In particular, bearing in mind the failures of the Ditsobotla Projects, what potential is there for the development of contract farming in ways that do not leave participants powerless and in debt? A recent survey of contract farming in Africa suggests that contract farming relationships are likely to become increasingly common, but draws mainly pessimistic conclusions about their impact. Significant income gains tend to be limited to participants with larger holdings, while many non-participant land-right holders are dispossessed or impoverished (Little 1994). Little suggests that there may be some positive potential for contract farming projects in which grower organisations lessen the power balance between producer and buyer, or where the buyer is itself a community-based organisation or NGO. Clearly, a crucial issue here is whether farmers' potentially greater involvement in decision-making would reduce the likelihood of their getting heavily into debt with the Project.

Thinking about the potential for extending commercial farming in Ditsobotla also raises questions about the social relations which shape access to resources and the ways in which resources are used. One important dimension of these relationships concerns access to land. At present, land cannot be bought and sold, on either Trust or Tribal land. Sharecroppers argue that they have little incentive to invest in the land they lease. For lessors, many of whom are women, the land they lease out provides at least a degree of security. The White Paper on Land Reform proposes an approach to tenure reform which recognises diversities in rights of tenure and which aims to create a legal mechanism for transforming current de facto relationships into formal legal rights to land. This seems a very positive development, although it will not in itself resolve the disagreements within communities over what tenure rights are appropriate.

One important implication of this discussion is that the scope for promoting livelihoods based purely, or even primarily, on farming, is probably rather limited in this region. This conclusion also suggests that the employment effects of an expansion of commercial farming would also be limited. Most rural dwellers in the region follow strategies involving multiple livelihoods (Lipton *et al.* 1996a, 1996b). Policies to promote rural livelihoods must recognise and encourage this diversity. They also cannot be taken in isolation from developments in the urban economy.

Land reform

In North West Province, there have been several thousand claims for land through restitution and 60 applications for land through the redistribution process. Fifteen of the latter were finalised in 1998. Now that the deadline for restitution claims has passed, DLA expects the number of applications for redistribution to increase (interview with Jeff Sebape, Deputy Director for Land Redistribution, North West Province DLA, Mafikeng, 4 March 1999). One group in Madibogo has applied for land through the redistribution process – the Siyaya Communal Property Association, which has applied for Portion 2 of the farm Maribogo. Maribogo farm lies on the south-eastern border of Madibogopan.[9]

The group was set up early in 1997. The members are mostly ex-farm workers from the Free State and the Transvaal. At first, there were 109 of them. However, many people became impatient with the slow progress of the application and membership fell to 37. Some new members then joined to bring the membership up to 76. Maribogo farm covers 352.53 hectares, of which about 300 hectares are arable land.

Maribogo is currently a mixed farming operation. Mr and Mrs Dry have owned the farm for the last forty years. They used to grow vegetables and fruit, but now grow only maize. They also have twenty-five dairy cattle. Six permanent workers and one domestic worker are employed on the farm. They used to operate a shop on the farm, but gave it up because of theft. Mr Dry also used to own a hotel and a shop in Madibogo. He still makes a supplementary income from transporting to the homes of farm workers grain they have received in payment for harvesting and stored in the silos at Madibogo. The Drys were never just farmers.

Theft has always been a problem, but they think it is becoming worse, 'because of hunger'. Also, while local people have always driven their stock on to the farm to graze, it is happening more often and people have started cutting the fences. The Drys are selling the farm for a number of reasons, including illness. Many other local farmers are selling, or thinking about selling.

The Siyaya group will receive a grant of R16,000 each to subsidise their purchase of the farm and pay for equipment, new infrastructure and repairs. As well as growing grain, they plan to produce irrigated vegetables, keep livestock and engage in broiler production. They also plan to re-open the shop. They intend to remain in their homes in Madibogo and travel to work on the farm. The group had received little preparation for running a co-operative business and dealing with issues of management and remuneration. The national DLA has delayed the application until the group is reduced in size and receives training.[10]

The fact that the current owners have depended on multiple sources of livelihood should raise concern about DLA's insistence that applicants for land through the redistribution process make a full-time commitment to the enterprise. Many applicants in the land reform process want to incorporate farming into multiple livelihoods. In a risky farming environment like that around Madibogo, such a wish is rational.

Multiple livelihoods: differentiation, risk and vulnerability[11]

Rural livelihoods, and the conditions which sustain them, are diverse, even within one region. Policy makers disagree about the scope for an expansion of smallholder agriculture. It is likely to be more feasible in regions of high-value land, close to markets, such as KwaZulu-Natal, than in more remote or semi-arid regions like Ditsobotla. In places like Madibogo, people will need to continue to engage in multiple livelihoods and the challenge is to find ways to support this.

Differentiation

The most significant inequality between households in this area, and one from which other differences follow, concerns access to a regular income. This may take the form of wages, remittances, or pensions. There are also, as would be expected, considerable differences in income levels between households receiving a regular income. However, households that do not receive a regular income have to follow quite different livelihood strategies from those that do. They are also significantly poorer. There are also significant differences in income and in livelihood-related activities between households with access both to land and to the means to work it and those – the large majority – who lack these resources.

Related to these inequalities, as both causes and effects, are inequalities in access to information networks (concerning, for example, information about jobs, or about how to get a pension), access to support networks, access to the state (for access to agricultural credit and state land) and the ability to enforce sharecropping contracts.

Recognising these inequalities suggests the following typology. Components of the typology should be thought of as positions between which households may move, rather than static groups. They are positions, rather than groups, because there is so much uncertainty and vulnerability in this risky environment. I discuss the most important of these risks below. Note that I am using households as units here. This is a rather crude approach, because of the complexities of movement in and out of households, but there is in each of these cases a core adult or adults living in the household permanently or visiting regularly. Membership/residence of others then fluctuates. This typology does not preclude looking at relationships within households and between them.

1   Households which have experienced income growth since the 1970s, or which have accumulated land, access to land or capital equipment (ten households). Three of these seem to be moving towards the second category.
2   Households whose income is relatively stable and/or which are managing from month to month (twenty-three households). Twelve of these households contain one or more members receiving a pension. Six households in this group seem to be moving towards the third category.
3   Households which are falling into greater poverty, which are obviously not coping (six households).

Households have moved between these groups in response to contingencies and to changing conditions in the local and national economies and because of changes associated with developmental cycles. There was significant movement from position 3 to position 2 during the 1960s and 1970s, particularly by the households of those ex-farm workers who arrived with very little and who got jobs in the late 1960s and 1970s. Some households struggled in position 3 until one or more members got their pensions. In the other direction, households have been affected by processes that limit their ability to find work, both structural and developmental cycle-related, or by processes that stem from the riskiness of their environment. Many commercial farmers are heavily in debt and unable to utilise their equipment fully. Many households in position 2 are vulnerable to falling into position 3. Oriel Mollootimile Mosadiwamakgoa's husband was a bus driver. In February 1999, he lost his job:

> 'He's looking outside for a job, in Itsoseng and Lichtenburg. It's very difficult to make a living at the moment. We keep on borrowing money and say that we will repay it when my husband gets a job. It's all uncertain. I borrowed money from other people in the sewing group and from a relative. I borrowed R50 from each of them. I can also go to my friends in Madibogo, but I haven't approached them yet. ... I don't know what to do now. I don't have money to look for a job and there are the children.'
> (Interview with Oriel Mollootimile Mosadiwamakgoa,
> Madibogopan, 29 March 1999)

Critical resources for being in position 1 are receipt of wage income at some point in the past, or now, which has made it possible to acquire fixed and working capital for farming, or access to wage income from a business partner. Critical resources for being in position 2 are a wage income, access to remittances or receipt of a state pension. If households lack access to these resources, rights over land which can be leased out in exchange for grain may partly compensate. A number of different processes may precipitate a fall into position 3, or may make it impossible for a household to move out of position 3. The most common are retrenchment or inability to find work, children's marriage (because children are not expected to remit much to their parents once they marry), or the arrival of new dependants (which may happen if offspring or other kin divorce and send their children back to be fostered). This last process increases the dependency ratio, while remittances from the child's parents may not be forthcoming. The dependency ratio may also be increased by the return of young, unmarried members of the household who have been unable to find work. Other factors which can push households into severe poverty are illness, accidents or the sudden death of a wage earner. All of these are common risks. Inability to get access to welfare payments may then prevent a household from coping with the consequences.

Vulnerabilities and household developmental cycles

It would be quite misleading to suggest that there is such thing as *a* household developmental cycle in Madibogo, let alone a Chayanovian developmental cycle related to the availability of farm labour. (See pp.127–8 for a discussion of Chayanov and household development cycles.) We have seen that there are profound differences between households in terms of the types of resources they have access to (land, livestock, wage income, more or less reliable remittances and pensions). So many households would look deviant in terms of a developmental cycle that one cannot identify a normal pattern – qualifications need to be made at every point. Nonetheless, there are points of decision, transition and strain which are systematically associated with household formation, maintenance and dissolution. In this respect, looking for developmental-cycle related processes is useful, not least because this forces us to distinguish these from other processes stemming from contingencies or structural changes in the environment.[12]

Household formation is often a drawn-out process, rather than a one-off event. Where a couple marry formally, a transfer of bridewealth from the man's family to the woman's family would be expected, though this transfer may be gradual. Until bridewealth is transferred, it is easier for either partner to leave a relationship and a young woman may continue to live with her parents even after the birth of children if bridewealth has not been paid. Additionally, when a woman comes to live at her husband's parents' home, she often lives and cooks with her mother-in-law for the first years of marriage. New households often form gradually, as husband and wife begin to provide for themselves and live apart from their parents-in-law, building their own house and finding and preparing their own supply of food. All this means that household formation is a process which depends upon there being resources available to cement a stable relationship and meet the costs of setting up a household. If these resources are lacking, it is very likely that young people will not be able to begin the process of forming a separate household. Moreover, because household formation is often a gradual process, it is reversible. Job loss or other contingencies can undermine it. In such cases, a woman may have a child without forming a household with the father. If she has a job, the child may stay with her parents and she may then give them some financial support. The father may or may not stay involved. If he can pay bridewealth, the woman may eventually come to stay with her parents-in-law, unless she has a job, in which case the children may stay with them. Once a couple has formed a household, whether or not they are formally married, there is a strong norm that their energies and resources should be concentrated on building up their own house, rather than supporting the previous generation. Resource transfers to parents tend to be greatly reduced. If the marriage breaks down, the children of the marriage may live with either partner's parents and these may then receive some support. The couple may be able to hope for support from their own children as they mature, provided that these children can find jobs. But they may themselves have to provide foster care for their grandchildren, for their siblings' children or for the children of more distant kin. As their own children marry, support drops off, just at the time when they may be unable to work. There may then be a period of quite considerable strain until they are able to get access to a pension.

In these processes of household formation, maintenance and ageing, vulnerabilities come from the prevalence of unemployment, insecure casual work and long periods of job search, the results of illness, accidents or violence and from the difficulties many women face in leaving their children to find work. Child fostering may alleviate these strains – twenty-two out of thirty-nine households were fostering children – as may labour migration on the part of the wife (sometimes leaving adolescent children alone), seasonal, casual work in agriculture, or access to money and food through social networks. The efficacy of such networks is limited by general poverty and by the tendency for some people to have more limited social networks than others.

Risks and vulnerability

The starting point for promoting sustainable livelihoods in this region must be an understanding of the sources of risk and the vulnerabilities of different kinds of household. It must involve thinking about ways to reduce the riskiness of livelihoods, reducing people's vulnerability to the effects of risk and avoiding increasing risk. Some of the most acute sources of risk stem from unemployment in the urban economy and the mining sector, but there are local sources, such as uncertainty about access to land, credit and sharecropping contracts.

Table 2.2 Major risks

All households	Commercial farmers
Retrenchment	Access to land (own land, sharecropped land and state land)
Long job-search	Ability to enforce contracts
Inability to find work	Access to capital
Inability to seek work	Access to appropriate, timely and affordable credit
Illness, disability or death of prime-age members	Rainfall
Weather-dependent demand for labour in agriculture	Prices of crops, inputs and livestock
Uncertain returns to trading or other informal sector activity	Crop theft
Spouse/children may not remit	Theft of inputs by workers
Marital failure within the household, for close relatives or offspring	
Arrival of new dependants	
For landholders: failure to find or sustain a satisfactory relationship with a farmer	
Livestock theft	
Death of livestock from drought	
Inability to access resources where the gatekeeper is the chief, or the state	

Responses

Risk explains people's concern to establish and maintain a rural base, to disperse their households and to diversify their sources of livelihood. In the case of farm workers, it explains their preference for low-waged seasonal work over the local informal sector – they get a year's supply of maize after the harvest. Employment in commercial farming is also risky (seasonalities, retrenchments), but it seems to look like a better bet than other options to many people.

What are the consequences of these processes for household structures? Mean household size in the households interviewed is five, with a range from one to eleven. The very large households are clustered around a regular income or a responsible adult. They have high dependency ratios, consisting mainly of older people with pensions fostering their grandchildren. The poorest households are young working-age adults who are unemployed. They may be too poor to look for work, unable to pay for transport costs and the bribes that are sometimes necessary to secure a job. They may be ill, disabled, or doing casual work. All but one of the most successful households are below average size, in terms of members living at home. This is because they are dispersed. In other words, one successful strategy is to have many children who are dispersed (provided that they can find work). The one successful large household is part of a larger network stretched across a 150-hectare farm in Geysdorp and a herd of 600 cattle in Kuruman. Ladlong Cornelius Masire, quoted at the beginning of this chapter, has a lease on state land at Geysdorp and his brother looks after the cattle at Kuruman. His father moves between his two sons, giving advice and making decisions with both of them. Ladlong also travels to Kuruman two or three times a month. Stretching multiple livelihoods over hundreds of kilometres is one response to risk. Other people depend on making minute calculations about the wages and food paid on different commercial farms, or on carefully dividing a pension between food purchases and working capital for trading.

Conclusion

The major problem people in this region face is the riskiness of their environment, in terms of climate, economy and social relationships. We have seen examples of households which have dealt with this highly risky environment successfully, but these are a small minority. It is doubtful whether, in such an environment, a significantly large number of households could emulate them. Past attempts to secure such an outcome have failed, creating a legacy of mistrust of government. While the new government retains a large degree of goodwill, it must be aware that failure in this area would have serious consequences for its rural support. The strategies people follow are not the result of conservatism or unwillingness to experiment. Rural households are resourceful and flexible, and the strategies they follow are sensible responses to risk. If the policy process makes the environment riskier, as the Drylands Projects did, then the policy itself is part of the problem.

Notes

1   In line with the policy of consolidating former RSA and bantustan districts into a unified local government structure, Ditsobotla District became part of the new Central District in 1996. Central District Council has jurisdiction over the old Magisterial Districts of Marico, Lichtenburg, Coligny and Delareyville in the former Transvaal Province and Molopo, Ditsobotla and Lehurutshe in former Bophuthatswana.
2   Bophuthatswana, Republic of 1993. Population data from the Bophuthatswana censuses are known to be inaccurate, as are data from censuses covering the African population of the RSA before the creation of the bantustans.
3   Stacey identified a group of commercial farmers selling more than ten tons of maize per annum to local depots, of whom he surveyed a hundred.
4   For a discussion of smallholder farming in Lehurutshe District, which is also in North West Province, see Drummond (1990).
5   Yawitch (1981) shows that many losers in this process lost land rights altogether.
6   These were later replaced by hired tractor-drivers.
7   See Drummond (1995) for an analysis of irrigation schemes in Lehurutshe.
8   For a discussion of the rapid repositioning of white farming co-operatives, see Amin and Bernstein (1996).

9   The following information comes from the following sources: interview with Jeff Sebape, Deputy Director for Land Redistribution, North West Province DLA, Mafikeng, 4 March 1999; DLA Memorandum Seeking Ministerial Approval for the Granting of a Subsidy for the Acquisition of Portion 2 of the farm Maribogo, Registration Division I.O. in the District of Geysdorp, North West Province; group interview with six committee members of the Siyaya Communal Property Association, Madibogo, 22 March 1999; interviews with Andrew Mokhutsane, Chair of Siyaya Communal Property Association, Madibogo, 24 March 1999, with Lydia Akanyang Tshatsha, Secretary of Siyaya Communal Property Association, Madibogo, 26 March 1999 and with Mr and Mrs Dry, current owners of the farm Maribogo, 26 May 1999.

10 Since this book went to press, the land reform programme has been suspended and its future remains undecided.

11 This section is based on forty interviews carried out in Madibogo, the nearby satellite village of Madibogopan and on state land leased by farmers at Geysdorp. Francis (1999) and Murray (1998) give details of research methods used in Madibogo and in the project overall, respectively.

12 Cf. Rahman and Hossain, writing about Bangladesh, who argue that the vulnerability of the poor can be understood in terms of a set of 'downward mobility pressures', namely: structural factors within the economy; crisis factors such as household contingencies and natural disaster and life-cycle factors, particularly the proportions of economically active and dependent persons in a household (Rahman and Hossain 1992, cited in Hulme and Mosley 1996).

3 Making a living
Migrancy and multiple livelihoods

Diversified livelihoods

Making a living in rural Africa today involves improvising responses to disappearing job prospects, falling agricultural output, collapsing infrastructure and the withdrawal of public services. What it does not involve is withdrawing from markets and concentrating on subsistence production. There is nowhere to retreat to, certainly not a mythical subsistence economy on the land. For most people, farming alone cannot provide them with an adequate living.

Rural Africans have been locked into markets since the colonial period: paying taxes, buying food, selling their crops and their labour. There are now too many things that people need money for to allow them to retreat into self-provisioning. Clothing has to be bought; children need school fees and uniforms. Consumption patterns have changed – people have come to expect to drink tea and use sugar and soap, even if they usually cannot afford to buy them.

Most compelling of all, most rural households do not grow enough food to provision themselves through the year and have not been able to do so for decades. Although this predicament is widespread and has many common features, it cannot be explained by a universal set of processes – 'overpopulation', 'urban bias', 'predatory states' or 'environmental degradation'. Rather, there are local histories of impoverishment (and accumulation), just as there are many different local presents, different predicaments, different ways of being locked into markets, and different responses. There may be as many differences within localities in this regard as between them. Differentiation *is* the norm.

One generalisation can be made, however. Evidence from across Sub-Saharan Africa suggests that rural people tend to construct their livelihoods by combining different income-earning activities. Farming often provides a surprisingly small proportion of rural households' incomes. Reardon reviewed twenty-three case studies distributed over Eastern, Western and Southern Africa and found that the average share of income earned in the non-farm sector was 45 per cent. It varied from 15 per cent to 93 per cent (Reardon 1997).[1] The share of non-farm in total income was particularly high in parts of Southern Africa, as case studies from Namibia (93 per cent), Lesotho (78 per cent),

Botswana (77 per cent) and the former homelands of South Africa (75 per cent) demonstrate (Sahn 1994; see also Ellis 1998). The poverty of farming in Koguta is underlined by the fact that the share there was 80 per cent. These findings corroborate an earlier review of fifteen surveys in nine countries by Sahn. They are also in line with a survey of rural localities in Ethiopia, Nigeria, Tanzania, Malawi, Zimbabwe and South Africa carried out in the mid 1990s, which found that the proportion of household income derived from non-farm sources ranged between 60 and 80 per cent (Bryceson 1999).

African rural livelihoods seem to be more diverse now than they were a few decades ago. When Haggblade, Hazell and Brown reviewed studies from the 1970s, they found less range in the shares of non-farm income. Some case studies also suggest that the share of non-farm income has grown (Haggblade *et al.* 1989, cited in Reardon 1997; see also Heyer 1996; Bryceson 1996; Ellis 1998; forthcoming). The Koguta case study also supports this hypothesis.

Diversified livelihoods may involve combining farming with wage labour, trading, selling services or producing commodities for sale. They also involve all the help, transfers, exchanges and information that people get access to through social networks. Some of these sources of income may be more dependable than others. They may also vary in the returns to labour that they offer and the resources people need to put into them. Just like commercial farming, not everyone has the means or capabilities needed to undertake them (Ellis 1998).

Researchers and policy makers concerned with rural poverty are now taking seriously the fact that diverse rural livelihoods call for different analytic approaches and policy interventions from approaches which assume that the main purpose of rural development is to promote smallholder farming (Berry 1993; Hart 1995; Lipton and Ravallion 1995; Scoones 1998; Carney 1998; Ellis 1998). Rather than concentrating their efforts on improving the productivity of small farms, policy makers are beginning to take as their starting point the need to support and enhance diverse livelihoods (Carney 1998; Ellis forthcoming). The challenge then becomes to understand how people construct livelihoods, what factors shape the strategies they follow. One approach being developed involves identifying three main elements – the assets that sustain livelihoods, the institutional context which shapes access and use of these assets and the livelihood strategies which these give rise to. The aim is to identify livelihoods that are sustainable (in terms of providing an adequate living, as well as environmentally) and determine how they can best be supported (Scoones 1998). Assets are defined broadly, to include social capital (networks, associational life) and human capital (education, skills and health), as well as natural capital (land, water, biological resources), physical capital and financial capital.

This approach does seem like a step forward. It starts from what people are actually doing to make a living and aims to support them in these efforts. It also recognises the important role that social resources play in livelihoods. It remains to be seen how difficult it may be to put into practice. One problem is that my sustainable livelihood may undermine yours, if I am appropriating land, for example, or if I shut you out of a trading network in order to reduce competition. Another is that social networks may be fallbacks, rather than adequate substitutes for other forms of capital. Strengthening social networks may in some circumstances be a poor substitute for more politically contentious action to redistribute access to other forms of capital. A third problem is that social capital itself can have a dark side (Putzel 1997). It may involve building trust among some groups at the expense of excluding others.

Diversified livelihoods may be a response to seasonalities in farm production (Chambers *et al.* 1981; Ellis 1998). While crop income may come in only at harvest time, households' consumption needs are year-round, often leading to a 'hungry season' before the harvest when food stocks, and cash income, are low. Food shortfalls are often most acute during the planting and weeding seasons, when demands on the household for farm labour are highest. Generalised shortages of food at this time mean that food prices are likely to be much higher immediately before the harvest (when food-short households are buying) than afterwards (when the same households may need to sell crops to raise money). Cash needs may also be seasonal. For example, school fees may be payable termly or annually. Having more than one source of income allows households to smooth income streams and deal more easily with calls on their incomes that do not coincide with seasonalities in crop production.

When households diversify their livelihoods, they are also likely to be trying to spread risks (Ellis 1998). Apart from a regular wage income, it is very likely that none of the activities from which people construct livelihoods can on their own provide a secure living. Many are likely to be risky and very dependent on the maintenance of local demand for goods and services. They give poor protection against generalised shocks to income, such as drought, or Structural Adjustment Programmes. In an uncertain environment, there is a strong incentive to spread risk and keep options open (Richards 1985; Berry 1993; Scoones *et al.* 1996). Having more than one source of livelihood means that the consequences of any one source failing will be less severe.

Contrasts between commercial farming regions and the rest should not be overstated. Commercial farmers also depend on income from other sources. Urban wages have played a central role in providing capital for farm investment across Africa (Kitching 1980; Haugerud 1981; Berry 1985, 1993; Low 1986; Collier and Lal 1986). This also means that households may diversify their livelihoods for quite different reasons. For some, diversification may be part of a strategy of accumulation. Some household members may seek wage employment or get involved in trading activities, while others may be more heavily involved in farming. This is why rural development policies which assume that smallholder households are farming full time do not fit African realities. However, many households in commercial farming areas do not grow the high-value crops. While some of them derive an income from agricultural labouring, others depend on non-farm sources of income. For them, diversification is a response to the inability of any one source of income to sustain them. What this means is that differences in livelihoods between more and less commercialised farming areas are differences in degree, rather than kind.

Migrancy and changes in urban economies

Labour migrancy has been the major link between rural households and markets in most parts of Eastern and Southern Africa for most of this century. Decades of labour migration have rested on households keeping footholds in both the urban labour market and a farming operation. In many regions, the backdrop to people's efforts to sustain livelihoods is the drying up of job prospects in the urban economy and the falling living standards of those who do manage to find work. Many formal sector workers in urban areas have lost their jobs and have had to turn to the informal sector to support themselves. At the same time, living costs in urban areas have continued to rise. New migrants have little prospect of finding a job, or of being able to remain in the urban area while they continue to look for one. In the past, new migrants could often rely on relatives who were already living in the urban areas to give them help with accommodation and contacts while they looked for work. Increasingly, urban dwellers are themselves so pressed that they cannot offer this support.

Some evidence to support these generalisations comes from Potts' longitudinal research in Harare. Potts carried out surveys of migrant households in Harare in 1985, 1988 and 1994. Macro-economic conditions were not as poor in Zimbabwe as in most African countries in the 1980s, since the economy continued to grow. In the 1990s, this growth disappeared. GNP did not grow at all between 1987 and 1997, while GNP per capita fell by 0.2 per cent per annum, making Zimbabwe's macro-economic conditions more similar to many other Sub-Saharan African countries (World Bank 1998, http://www.world-bank.org – country data). Additionally, Zimbabwe went through an Economic Structural Adjustment Programme in the early 1990s which lowered formal-sector employment and raised the cost of living in the urban areas. Potts found an increasing trend for the heads of such migrant households to be in work, rather than seeking work. In 1994, 96 per cent of the migrant household heads were working. The reason why so few household heads in Potts' sample were seeking work was almost certainly that such people have found it financially impossible to remain in town and have moved back to the rural areas. Overall, she found that unemployment was squeezing people out of Harare more quickly and easily than in the 1980s and that urban incomes were becoming more and more inadequate (Potts 1999).

Understanding 'hidden' livelihoods

The improvised responses of many rural households to poor urban job prospects often still depend on maintaining a toehold in the urban economy, whether through trading or through the oscillation of some of its members between the urban informal sector and the countryside. People, goods and money move between city and countryside in complex networks of market and non-market exchange, much of it 'hidden' from official gaze. In some regions, migration may be increasing, as young people fail to find work locally and move around in search of work in commercialised farming regions or in small mines, as Madulu found in North West Tanzania (Madulu 1998). Complex trading networks have also emerged, as people try to take advantage of regional differences in supply and demand for different commodities. Cross-border differences in regulation enhance opportunities for such networks to flourish.

The analytic tools for understanding these 'hidden livelihoods' have been developed mainly with the urban informal sector in mind. Indeed, policy makers have often overlooked the income-earning activities of people in rural areas when they have attempted to characterise the informal sector (King 1996; Allen 1998. Studies of the rural informal sector in Africa include Chuta and Liedholm 1990; Liedholm and Mead 1993). The informal sector in Africa was 'discovered' in the early 1970s, influenced by the ground-breaking work of Keith Hart on informal income-earning activities in urban Ghana and then taken up by the 1971 ILO (International Labour Organisation) Employment Mission to Kenya in its influential study of micro-enterprises (Hart 1973, 1992; ILO 1972). Since then, it has become increasingly clear that 'hidden livelihoods' are central to the efforts of ordinary Africans to make a living. They are also the realm within which the privileged make use of the networks and cultural repertoires that allow them to accumulate wealth (MacGaffey *et al.* 1991; Bayart *et al.* 1999).

Early attempts to characterise the informal sector saw it as a residual, a response to restricted demand for labour in the formal sector in a dual labour market. Neo-Marxists argued that this approach ignored the links between the formal and informal sectors, stressing instead that involvement in the informal sector is a form of disguised unemployment, while the cheap products that petty commodity producers sell to the urban poor provide a subsidy to formal-sector employers (see, for example, Portes and Walton 1981). More recent approaches emphasise the agency of participants in the informal sector. Informal sector metalworkers in Kenya display great ingenuity and technological confidence (King 1996). Tanzanian micro-entrepreneurs developed their 'projects' in resistance to the tight state regulation of economic activity in the 1980s. In doing so, they successfully challenged the state's regulatory framework, forcing the Government to recognise and accommodate the growth of private-sector activity (Tripp 1997). Tripp argues that this internal pressure to liberalise state regulation of the Tanzanian economy is often overlooked in accounts of structural adjustment that look only at external pressure from donors.

Tripp also found that the informal economy was highly socially embedded. Micro-entrepreneurs relied heavily on social networks for customers, credit and savings, market information and contacts. These networks were built around kinship or quasi-kinship relations constructed with non-kin.

The embeddedness of market activity in Africa has been evident for a long time.[2] Cohen found that the Hausa monopoly of long-distance trade in kola nuts and cattle between northern and southern Nigeria was made possible by the development and active maintenance of ethnic exclusiveness by Hausa traders in the south (Cohen 1969). Marris and Somerset stressed the centrality of kinship to the organisation of African businesses in Kenya (Marris and Somerset 1971). Pottier's study of Mambwe cross-border trading networks (discussed later in this chapter) shows how Zambian villagers relied on kin across the Tanzanian border to facilitate their trading activities.

The persistence of embeddedness can be a great source of strength and resilience to people improvising a livelihood. It is often easier to resort to help from kin than non-kin as a fallback. Kinship-based networks can be a valuable source of information. It may also be easier to build relations of trust with kin, or quasi-kin, than with non-kin. The kinship mode of organisation is a 'cultural repertoire' that constitutes a form of social capital in African societies (Bayart *et al.* 1999). But relying on kin or kin-like relations can also be a hindrance to accumulation and to management of an enterprise along profit-maximising lines, as Marris and Somerset argued for small businessmen in Kenya and Berry later showed in her research on motor mechanics in Western Nigeria (Berry 1985). It is hard to discipline or fire someone who is kin.

Studies of the informal sector show that it is highly differentiated, by income, location and gender and in terms of profitability and links to the formal sector (Scott 1994; see also Seppälä 1996, discussed later in this chapter). A major analytic challenge is to understand how the growing importance of 'hidden livelihoods' in the African countryside feeds into processes of accumulation and impoverishment, how it impacts on processes of differentiation by locality, class and gender.

One starting point for this is to identify the different types of 'livelihood systems' that people engage in. Grown and Sebstad (1989) suggest that the dynamics of poverty should be approached through a 'livelihoods systems' framework.[3] A livelihoods system is:

> the mix of individual and household survival strategies, developed over a given period of time, that seeks to mobilize available resources and opportunities. Resources can be physical assets, such as property, human assets such as time and skills, social assets, and collective assets like common property (forests) or public sector entitlements. Opportunities include kin and friendship networks, institutional mechanisms, organizational and group membership and partnership relations.
>
> (1989: 941)

Just as more recent work on diversified livelihoods stresses, Grown and Sebstad argue that livelihood strategies involve combining many different activities and relationships. These may include involvement in labour markets, savings, accumulation and investment, but also social networking and income, labour and asset pooling. People adjust the mix of activities according to the season, locale and climate and according to their age, position in the life cycle, educational level and time-specific tasks.

One important issue is whether it is appropriate to conceptualise what people are doing in terms of 'strategies', rather than seeing what they do as responses to the constraints of their circumstances. When does the term 'strategies' over-state people's ability to act as agents, rather than having to react as victims of circumstances (Rakodi 1991, cited in Beall and Kanji, 1998)? How much room for strategising do the poor have? Ellis argues that there has been confusion in the literature between household strategies for spreading risk (which are active) and coping strategies, which are essentially reactive. He makes a useful distinc-tion between *ex ante* risk management and *ex post* coping with crisis. Reardon also argues that income diversification may be carried out *ex ante*, to reduce income risk, or *ex post*, in response to low farm productivity and income shocks. It may also be undertaken to earn the money needed to finance farm invest-ments (Ellis 1998; Reardon 1997; see also Carter 1997; Kinsey *et al.* 1998).

Sometimes people have no option but to react to the pressures of circum-stances. At other times they are able to be active agents, even perhaps changing those circumstances, or moving out of them. Analyses of livelihoods need to distinguish between explicit strategies on the one hand and reactions to systems of power and situations of crisis on the other (Beall and Kanji 1998). Making this distinction is often hard, however, because it can be difficult to assess how far people are able to act strategically.[4]

The goals of these strategies are also likely to vary. Grown and Sebstad suggest that the goal of the poorest groups is survival; that the goal of people whose basic survival is assured may shift to security and that people who have achieved basic security may pursue growth. They argue that the shift from survival to security is marked by a diversification of the livelihoods mix. The shift to growth may be characterised by a concentration of investments on higher return but riskier commercial enterprises. This typology provides a useful starting point for understanding differences in livelihoods, although the distinc-tions it makes need to be investigated, rather than assumed. In some circumstances very poor people may have to rely on a mix of activities, precisely because none of them are reliable. However, there is African evidence to support the suggestion that the livelihoods of the poorest may be less diverse than those of the less poor. In the studies reviewed by Reardon, upper income strata households had much higher shares of non-farm income in total income, as well as higher absolute non-farm incomes. Inequalities in access to non-farm income are often a major part of the explanation of rural inequality (Reardon 1997). However, the fact that households derive a substantial proportion of their income from a non-farm source like a pension or remittances does not mean that their livelihoods are necessarily more diversified than other house-holds. In the Koguta case study, a number of better-off households consisted of women and children who derived most of their income from the husband's remittances. Their farm incomes were small, but they also did not feel it was either necessary or appropriate to earn an income locally. Such households seemed to be increasingly unusual, however, as migrants' ability to send home adequate remittances was declining.

The livelihood systems framework shows that policy makers should avoid assuming that people always pursue growth strategies, or that their efforts are concentrated on particular enterprises. As we have seen, in much of rural Africa, diversified livelihoods are a sign of growing insecurity. Many would-be accumulators are hampered by the need to stay diversified.

At first sight, many local economies in rural Africa appear to be flourishing. Markets are full of traders and buses and taxis are crammed with people, livestock and produce. But the level of activity in markets is often more a sign of poverty than of economic dynamism. Overall demand is constrained by low incomes. Many of the commodities being sold (groceries, pots and pans, containers) may be brought in from other regions or from overseas. Local producers of textiles and clothing may be out-competed on price and quality by mass-produced goods. Agricultural produce may be brought in and re-sold from other, more prosperous regions. The growing importance of non-farm activities in rural livelihoods is a sign of the long-term decline of agricultural production in many regions. 'Deagrarianisation' is observable across Sub-Saharan Africa (Bryceson 1996). Additionally, most rural households can no longer rely on adequate remittances from migrant relatives.

What implications do these changes have for rural households? Are livelihoods built around non-farm activities different from livelihoods based mostly on farming and livestock production? Is rural differentiation around non-farm activities different from differentiation around farming? How do livelihoods based on non-farm rural activities differ in their implications for households from livelihoods that rely on migrants' remittances?

Just as with farming and livestock-keeping, households and individuals differ greatly in their ability to engage in non-farm activities and a rural economy in which most livelihoods rely on these activities for an income is likely to be highly differentiated. The bases of differentiation are likely to bear some resemblance to differences in a more strongly agricultural local economy. Access to household labour will be important. Larger households are more able to diversify their portfolios of activities. Households with access to remittance incomes are more likely to have resources to reinvest in non-farm activities, although they may also have a less pressing need to undertake them. The kinds of activities people undertake and their ability to earn an income from them are gendered, as well as being strongly affected by households' positions in developmental cycles (Bigsten and Kayizzi-Mugera 1995; this is also shown in the Koguta case study). People's ability to engage in non-farm activities will often depend on access to information and social networks. Access to networks also, of course, determines their ability to make claims for assistance. Spatial advantages (proximity to a road, or to an international border, for example) become more important and land resources less important in this type of economy (Booth *et al.* 1993). Socio-cultural differences may also be important, shaping people's access to material resources (cf. Seppälä 1996, below).

Two studies of former migrant labour economies in Northern Zambia explore how rural people have responded to labour migrancy and its demise.

Figure 3.1 Zambia and Zimbabwe

Zambia's Northern Province: migrancy and commercial farming

Northern Province is a poor, marginal region in North Eastern Zambia. As in many other parts of Southern Africa, its economic history has been shaped by the rise and demise of labour migrancy. Migration began in the early years of the century, to Katanga (Congo), Southern Rhodesia and South Africa, later to the Zambian Copperbelt, to mines and sisal plantations in Tanganyika and to jobs within the Province. Copperbelt migration intensified in 1930s as copper production expanded and copper became the mainstay of the national economy (Richards 1939[5]). It is not possible to give an accurate estimate of the numbers of people involved in migrancy, as colonial population and migration statistics are notoriously unreliable (Moore and Vaughan 1994). Migration rates also varied greatly between different parts of the Province. Moore and Vaughan, in their study of agrarian change among the Bemba people, quote estimates of male absenteeism from different areas in the late 1930s and early 1940s ranging from 40 to 75 per cent (Moore and Vaughan 1994: 147–8). Contemporary assessments from this period show a great deal of anxiety about what were believed to be the disruptive effects of this large-scale migrancy. The discussions of colonial officials and anthropological writing about the region were characterised by a pervasive sense of breakdown. However, the colonial government made little effort to promote increased production in agriculture and made little investment in infrastructure (Richards 1939; Moore and Vaughan 1994). Throughout the colonial period, Northern Province was essentially a migrant-labour reserve.

Colonial officials' anxieties revolved around what they saw as the related issues of residential mobility, shifting cultivation and food insecurity. In the early 1930s, Audrey Richards carried out pioneering research on food production and nutrition among the Bemba. She found the Bemba diet to be seriously inadequate in terms of the calories it provided. It was subject to great variation, with a prolonged annual hungry season. She also argued that the food production system was becoming more inadequate. Richards identified two kinds of reasons for this development. One was the loss of the labour of male migrants at critical points in the agricultural cycle. The other set of reasons resulted from the links between the political system, Bemba identity and agricultural production. Richards argued that the centralised, pre-colonial Bemba political system gave ritual and organisational cohesion to the cultivation system. Chiefs marked successive phases of the agricultural cycle through rituals of tree-cutting, crop planting, first fruits and so on. Rituals gave a sense of cohesion, as well as providing an impetus for productive activity. Richards thought that the weakening of chiefly authority diluted this co-ordinating role and thus contributed to the decline of agricultural production. Richards also placed great stress on the values and cultural practices which she believed contributed to the poor state of Bemba agriculture. She suggested that the warrior tradition of Bemba men devalued the importance of agriculture in their eyes (Richards 1939). Moore and Vaughan have deconstructed this analysis, arguing that its strongly culturalist element reflects colonial discourses which linked ethnic identity with mode of livelihood. They also question whether chiefly authority had ever been as great as Richards believed (Moore and Vaughan 1994).

During the colonial period, Bemba people lived in scattered settlements. These settlements were often residentially unstable, for a number of reasons. Essentially, residence rules were flexible enough to allow people to keep their options open. Widowed, divorced or abandoned women could go to live with various members of their matrikin. When male migrants returned to the rural area, their place of residence was, similarly, open to negotiation (Moore and Vaughan 1994: 169–70). Bemba farming systems involved shifting cultivation (*citemene*) of millet, together with cultivation in more permanent gardens (*ibala*). Colonial officials worried about endemic food insecurity in the region and wished to develop agriculture enough to make labour migration a less attractive option. If the Bemba would only live in more stable settlements and settle down to permanent cultivation, a progressive peasantry would emerge and agricultural development would be possible.

Moore and Vaughan argue that, in fact, Bemba farming systems and household economies have proved far more resilient than observers believed, precisely because the negotiability of social relations has allowed people to keep their options open. Bemba people have faced a set of recurring problems through the colonial and post-colonial periods. These include the problems of how to control labour; how to maintain social networks and how to reproduce households. The solutions people have used have also been recursive, drawing on old repertoires of solutions (such as residential mobility) in order to develop new ones.

Northern Province has experienced far-reaching economic changes in recent years, as demand for labour in the mining sector and the urban economy has collapsed. It is estimated that the proportion of the Zambian labour force in formal-sector employment fell from 16 to 10 per cent between 1975 and 1990, while average real earnings in the formal sector had fallen to 29 per cent of their 1975 level by 1990. The large majority of urban households now have to rely on the informal sector for their livelihoods (Loxley 1995). All this means that labour migration no longer looks like a viable option to many young people in Northern Province, while many long-time urban residents have returned to the rural areas, because they can see that they have no prospect of being able to live on in town after retirement. Now that Northern Province is an ex migrant-labour economy, people have to make a living locally. This transition is becoming increasingly common across the region. Moore and Vaughan believe that the demise of labour migrancy has simply changed the form in which the recurring problems of labour control, the maintenance of social networks and the reproduction of households present themselves.

However, these responses now appear to be standing in the way of a large-scale expansion of commercial farming. In the 1980s, the Zambian Government promoted a 'back to the land' policy, hoping to encourage smallholder farmers to take on commercial production of hybrid maize. Unlike earlier efforts to develop commercial agriculture in the region, this attempt seems to have been relatively successful. Between 1975 and 1988, production of maize in Northern Province rose by around 850 per cent. This happened despite delays in the delivery of inputs, the deterioration of the transport infrastructure, delays in payments and shortages of the consumer goods which crop earnings could be used to buy. The expansion probably took place because the collapse in urban employment changed calculations of the relative attractiveness of farming and wage labour. Most of the more commercialised farmers were returned migrants. Their earnings provided the capital needed to grow hybrid maize.

For most farmers, however, their income from farming did not cover their cash needs. Even if it did, women often did not have access to this income. Agricultural income tended to come in a one-off payment and was used for bulky expenses like clothing. This made off-farm activities vitally important for financing day-to-day expenditures. Money was also crucial for getting access to extra-household labour for the farm. Moore and Vaughan's women informants made a direct link between income generation, investment in labour and house-hold reproduction.

> Thirty women were asked about the relationship between their farm and off-farm activities. Twelve brewed beer as a way of gaining access to labour. Four of these gave their labourers beer at the end of the day. The rest exchanged the beer for cash which they used to buy commodities. These were used for household consumption or as payment for their labourers. Seven women engaged in petty trading either to raise money to pay labourers or the commodities they needed to pay labourers or to consume at home. Five women worked for other women for food, clothes, sugar, salt and fish. They sometimes exchanged these for other commodities, or to pay their own labourers.
>
> (Moore and Vaughan 1994: 227–8)

Maize cropping has been associated with changes in patterns of land use, in the field system and in the gender division of labour in agriculture. Adoption of hybrid maize increased *ibala* cultivation at the expense of *citemene* cultivation and this change has been associated with greater male control over household labour and the crop income (Moore and Vaughan 1994: 216). Hybrid maize is a labour-intensive crop and male household heads seemed to be seeking greater control over the labour power of other household members. This increased women's labour time and, presumably, reduced their ability to work either on their own account on their gardens, or to engage in off-farm activity. Such a change would make households less able to keep their income-earning options open. Even with commercial maize farming, households in this region displayed resilience, rather than any real prosperity, and this resilience had for a long time rested on their adaptability. Ironically, maize farming may have made house-holds less adaptable and resilient.

When migrants returned, then, the focus of gender conflict might shift from access to remittances to control over labour time and cash income. Women complained that they often had to leave work on their own fields undone in order to work on their husbands' hybrid maize crop, especially when it came to weeding, or that they grew smaller amounts of beans, groundnuts, cassava and relish crops than they would have wished (Moore and Vaughan 1994: 219–20). The trend towards conflict over access to labour and over income emerges clearly in the case studies of gender and rural livelihoods discussed in Chapter 4 and it was often a source of bitter conflict in Koguta. The Bemba case also shows that policy makers intent on reconstructing smallholder farming in former migrant labour economies should be less sanguine about the impact of such a change on power relations within households, on women's access to land and their control over their labour time and money incomes (cf. Delius 1996: 222, commenting on South African policy debates about the former bantustans). Men's attempts to capture women's labour time may make it more diffi-cult for women to keep their livelihoods as diversified as they might wish. This case shows that the issue of whether a household should diversify its livelihood or not may generate conflict over access to labour. The term 'household strategy' would be highly misleading in this kind of situation. The 'strategy' is the outcome of conflict between household members.

In other parts of Northern Zambia, the Government's efforts to promote commercial maize farming were much less successful. In North West Province, some farmers expanded maize production in the late 1980s in response to government provision of subsidised inputs and support for marketing. However, the Government's turn towards market-based policies in the 1990s led to the withdrawal of these supports. The small boom in maize production promptly collapsed (Crehan 1997).

After migrancy in Mambweland

Mambwe territory lies in North Eastern Zambia, adjacent to the region studied by Moore and Vaughan, and to the border with Tanzania. In the late 1970s, Pottier found that people living in Mambwe territory, lacking access to jobs in the urban economy, had attempted to reactivate the flow of cash into the region. They constructed trading networks with kin living across the border with Tanzania and began growing beans for sale. Mambwe areas had experi-enced historically high rates of labour migration. In 1953, Watson's census of eight villages found that 42 per cent of men aged between fifteen and forty-nine were absent, rising to 66 per cent of twenty to twenty-four year olds. Migrants were working in Tanganyika, in the Copperbelt and other towns, or elsewhere in Northern Province (Watson 1958: 61). Watson argued that high rates of migration were compatible with the maintenance of agricultural production in Mambwe villages because they were residentially stable.[6]

The context to Pottier's study is the virtual ending of labour migration to urban areas. Formal employment opportunities had virtually disappeared by the late 1970s. It was also increasingly difficult for new urban migrants to establish themselves both residentially (staying with urban relatives) and in informal sector trading. Long-term urban residents were anxious to preserve their own access to livelihoods and not lose them to incomers. Many migrants had returned to Northern Province, but were currently living in small towns, such as Masaiti, the informal settlement around Mbala. Pottier suggests that people were undertaking a stepped process towards an eventual return to village life at an opportune moment. Living in Masaiti allowed them to retain an urban lifestyle.

There are parallels here with strategies for returning to the countryside that Tripp found in Dar es Salaam. In the 1960s, observers of urban residents in Dar es Salaam noted the declining importance of income from agriculture and concluded that workers were likely to remain in the city after retirement (Von Freyhold 1972; Market Research Ltd 1965; both cited in Tripp 1997). The 'classic' African pattern of circulatory migration seemed to be giving way to more permanent rural-urban movement. By the late 1980s, all this had changed quite dramatically. Not only did many urban residents intend to retire to the rural areas, but many were leaving their jobs and moving to the countryside to farm. It was becoming extremely difficult for people living in Dar es Salaam to survive in the face of massive falls in real wages and many now believed that farming held more promise than life in the city. Large numbers of city residents grew crops to supplement their meagre salaries. Eighty-three per cent of Tripp's sample of residents in Manzese, an area populated mainly by workers and small-scale entrepreneurs, had acquired plots in Dar es Salaam itself or on its outskirts. Almost half farmed in their home areas, which for two-thirds was the city and its vicinity. Many workers were attempting to acquire enough capital to move permanently to the countryside.

Many people living in Masaiti were involved in informal trading, mainly in agricultural produce. Masaiti was the hub of a complex trading network covering the Northern plateau and the Tanganyika lakeside, much of it cross-border. Many people living in villages were also heavily involved in (illegal) cross-border trading. However, all this activity was not a sign of economic buoyancy, but a function of the region's proximity to the border with Tanzania. The border created a situation of economic complementarity: it was a stimulant for trading, not a line of separation. There were shortages of essential goods on both sides. Tanzanians wanted consumer goods from Zambia, while Zambians wanted agricultural produce from Tanzania. These options were opened up by differences in states' policies and by the availability of kinship as a resource for underpinning trading. People had kinship ties on either side that they drew on to support cross-border trading, getting both information and logistical support from kin.

In Chivunzila village, Pottier found that Tanzanian trade partners were often connected through women's kinship links. Many Chivunzila women were born in Tanzania and so had close friends and relatives across the border. The mother-daughter link was the one most frequently activated for trading and women were heavily involved, though it was risky and the travelling involved made it arduous. The kinship link gave Chivunzila women a right and means to claim any profits their menfolk made in cross-border trade. The village's location so close to the border also encouraged women to return there on divorce and to remain there after re-marriage. Kinship and the border provided some women with a niche in the local economy. Pottier suggests that this example illustrates a more general point: there was a growing flexibility in the residential make-up of Mambwe villages, as people moved around to find niches in an increasingly impoverished rural economy. This represented a change from Watson's picture of residential stability. It was much less likely to be compatible with maintaining agricultural production.

Three processes had led to a growing shortage of land. The demise of labour migration, a national ban on moving village sites and the government's (now abandoned) attempts to enforce village regroupment increased pressure on land

and reduced soil fertility. Intensive cultivation of crops for sale now threatened ecological damage. Increasing population density had led to adoption of new methods of cultivation and new crops (cassava, hybrid maize and commercial production of beans). The major government intervention in the 1970s, as in Bemba villages, was a programme to encourage hybrid maize production, in a conventional development package of credit, inputs and technical advice. Pottier found that returned migrant farmers who had taken up hybrid maize production had found it disappointing. The crop was ecologically unsuitable. The soil was too acid and heavily leached and the rainy season was too long for maize cultivation. There was also widespread antagonism to the government's developmental organisations and suspicion of AFC (Agricultural Finance Corporation) loans, which were thought to involve taking on too much risk. 'Emergent farmers' were generally thought to be better off than average, but were plagued by low prices, uncooperative extension workers and irregular supplies of inputs. Since trading was also very risky, many men and women began growing beans for sale. Bean growing and cross-border trading were perceived as the best strategy for avoiding impoverishment, but the general picture was one of growing poverty.

Differentiation had increased since the 1950s, as the group of people needing to sell their labour to meet basic needs had grown. Many poor people worked on others' farms in exchange for food and cash. The apparent continuity here with older, reciprocal exchanges of labour through work parties was illusory. They systematically benefited the better off, who received more labour than they donated. In other villages, wage labour had become the norm. Social antagonisms were defused, however, because there were still loopholes left for the disadvantaged. Trading networks of agnatically related women allowed for a reasonably adequate supply of food and cash in the face of falling production. They allowed people to compensate for declining levels of production by activating kinship links, but trading networks could be sustainable only up to the point where there was still something to be exchanged (Pottier 1988: 18).

Mambwe villagers were resourceful in the face of the collapse of migrancy, but the niche they had found was very insecure, depending on the vagaries of demand in Tanzania and their continued ability to avoid border controls. In the Bemba case, the fluidity of the household's composition and residence and the flexibility with which they could call on kinship relations were vitally important to their ability to diversify. Pottier thinks that kinship relations among the Mambwe were becoming more fluid, in response to similar pressures.

So there are points of similarity and contrast between the two regions. They were both areas of high out-migration for many years. From the 1970s onwards they were confronted by a major change in their links to the wider national economy. Labour migration is no longer an option for young Bemba and Mambwe and many older people have retired back to the area after decades of urban living. The Zambian government's promotion of hybrid maize seems to have been more successful in Bemba areas, though Moore and Vaughan could see little to explain why the most recent attempt had been more successful than earlier ones. Even so, in both areas the large majority of households cannot survive through farming alone. They need to look beyond the farm for ways to meet their needs for cash. People try to meet those needs through diversifying their sources of income and keeping their options open in environments that are increasingly risky and uncertain. The next case study deals with the issue of risk explicitly.

Chivi District, Southern Zimbabwe: responding to risk and uncertainty

After Zimbabwe's independence in 1980, the Government followed a dual policy towards the agricultural sector. The first component of this involved some land redistribution from the white-owned large farm sector, in the form of both the transfer intact of large farms and the subdivision of farms between Africans through resettlement schemes. The second component has been to try to improve services and raise production in the communal areas (the former African reserves). There are, therefore, four sub-sectors: the remaining large farms, resettlement areas, areas of small-scale commercial farming set up in the colonial period and communal areas (Zinyama 1995). The communal farming sector supports about 51 per cent of the population. Migration from communal areas to the towns is widespread, but, as is so common elsewhere in Africa, migrants try to maintain access to rural land and many households pursue a dual strategy of farming and urban employment. Holding on to land also offers urban workers the prospect of some security in their old age (Potts and Mutambirwa 1990; Zinyama 1995). Output from smallholder farms has expanded, but poverty and malnutrition are widespread in the communal areas (Christensen and Stack 1992). They are particularly common in the drier, southern and western communal areas, which are subject to recurrent droughts, growing population pressure and land degradation. In these areas, rural households are very dependent on grain production and incomes tend to be low and highly skewed (Jackson and Collier 1988).

Chivi Communal Area lies in Southern Zimbabwe. Scoones and his colleagues carried out research into livelihoods in Chivi in the early 1990s (Scoones *et al.* 1996). They intended to focus on household responses to an environment which is generally full of uncertainty, but found themselves observing one of the most severe droughts of the century in 1991–92. Their study of livelihood strategies therefore includes a study of the coping strategies followed in response to this drought. Corbett carried out research in Shindi, Chivi District, in the same period, with a focus on livelihoods, food security and nutrition (Corbett 1994).

In order of importance, sources of cash potentially available to people in Chivi are remittances from wage labour; money from sale of crops and livestock and local piecework, crafts, manufacturing and trading. Money comes from these various sources in differing rhythms and is used for quite different purposes. Farming in Chivi is a risky affair, while all the other sources of income are beset by uncertainty. In most of the years following Independence, agricultural production has been too low to allow farmers to sell crops and many farmers did not sell even in the good years. People rarely sell cattle and when they do it is normally in response to an immediate need for money. Crop and cattle sales are used to finance lumpy investments. Other forms of income come into households in smaller, sometimes regular amounts and are used to service smaller, recurrent expenditures. Wage labour and remittances are becoming less and less reliable sources of money. Real wages have declined since the mid 1980s, particularly for people in low-paid jobs, and employment is much less secure. Price inflation has reduced the value of remittances (Scoones *et al.* 1996: 54–5). People in Chivi are involved in many other income-earning activities, such as trading, beer brewing, farm labouring, building work and hiring out ploughs and carts. These activities are differentiated by gender, some depend on possession of special skills or resources. Some, like selling crafts or trading, are more lucrative for people living closer to main roads or to business centres.

Scoones *et al.* looked at the strategies individuals and households follow to construct livelihoods in the light of the opportunities that they have and the uncertainties that they face. They looked at returns to labour in common income-earning activities and the trade-offs that these imply. Some activities are very dependent on climate, others are more affected by the distribution of skills, while others are shaped by macro-economic conditions. Returns to labour in agriculture are low. While investment in livestock offers higher returns, the danger of drought makes this risky. Most local non-farm activities generally provide low returns to labour, but they may be relatively stable sources of income. The activity with the highest and most dependable returns is formal employment. However, as we have seen, this is becoming harder to find and hold on to.

People's response to these uncertainties is to construct a portfolio of income sources, from farming and non-farm activities. Scoones *et al.* found that the need to construct a comprehensive portfolio gives a clear advantage to larger households, even though these also have more mouths to feed (Scoones *et al.* 1996: 66). They also found that some people followed more individualised strategies for survival and accumulation when times were more plentiful and they could do this without jeopardising the survival of others. On the other hand, households also often pooled their resources and labour with other members of their kin groups in joint activities like beer brewing. What they also stress is that people have to be flexible in the ways in which they construct their portfolios, responding to changes in the environment from season to season.

In order to understand responses to the 1991–92 drought, Scoones *et al.* make a distinction between two components of livelihoods. The first, production-storage-consumption, failed almost completely in 1991–92. The second component of livelihoods involves other assets, investments, markets and claims. It includes households' exchange entitlements, through their involvement in markets and through their participation in non-market exchanges of labour. It also includes participation in non-market exchanges of labour and their ability to make claims of support and assistance. Local networks of support were important for the vulnerable. Labour and draught-animal sharing clusters were common and centred around close kin relationships. Food sharing within clusters went up during the drought, as did borrowing of money.

In a series of ten case studies, Scoones *et al.* look at how people combined sources of livelihood in different strategies. Two cases, from wealthier and poorer households, show how resourceful rural households have to be in shifting their strategies to respond to changing circumstances.

In Mr H.'s home, household members earned money from poultry production, from selling livestock, from beer brewing, vegetable production and sale of skins and hides. The father also attended a government-run food-for-work programme. Members of the household did not carry out these activities all at the same time. Rather, they shifted between them during the different phases of the drought. Early in the year, the household sold livestock. This became more difficult as their condition worsened. Mr H. then thought it better to slaughter the animals and eat the meat. Mrs H. brewed beer early on in the drought, but this became impossible when the grain ran out. Instead, they grew vegetables and Mr H.'s son and daughter, both teachers, increased their remittances. The household sold poultry throughout the drought, which provided cash to buy food and feed for the livestock.

Mr D. comes from another part of Chivi and had no relatives to assist him during the drought. Finding it difficult to feed a family of twelve, he went gold panning at the Runde river. His family also relied on government food aid, but this source was unreliable. Because gold prices were low later in the year he stopped gold panning and sold three of his five donkeys to buy food. He also sold one of his remaining goats.

(Scoones *et al.* 1996: 186–7)

Different strategies varied in their reliability as bases of livelihoods. The success of commodity exchanges depended on supply and demand conditions for the commodity, on the availability of labour and, often, on the availability of capital. In 1991–92, the most successful exchanges were those relying on income coming in from outside Chivi. Significantly, people used networks of kinship, friendship and past contacts in order to secure exchange entitlements, as well as to make claims directly. So it is misleading to draw a sharp distinction between market and non-market forms of access to resources.

People's ability to follow different livelihood strategies and their inclination to do so, is shaped by many considerations. Their endowments of key resources – land, labour, livestock and money – limit the options they face, while their decisions will also be shaped by the returns to their labour that different activities offer. Women's decisions will also be affected by the extent to which different activities are compatible with domestic labour, by their ability to negotiate use of their labour time in different activities and by their prospects of benefiting from the proceeds of different economic activities (see Chapters 4 and 8). Men may also find it easier to keep control of the proceeds to some activities than others. Ferguson found that Basotho miners preferred to invest their wages in livestock than in farming. Livestock purchases were construed as long-term investments in 'building the house', helping men to resist pressures from their wives to sell livestock to meet their need for cash (Ferguson 1992). Differentiation around livelihoods may also reflect socio-cultural and political divisions, as Seppälä found in South Eastern Tanzania.

Lindi District, South Eastern Tanzania: income generation and cultural spheres

Kilimahewa village lies in Lindi District, a poor, peripheral area in South Eastern Tanzania. The local economy is heavily dependent on growing coconuts and cashews. It is a diverse and divided village, with quite substantial differentiation between households around access to better-quality land (suitable for growing coconuts) and access to labour. There are also distinct subcultures in the village, with divisions around religion, generation and politics all shaping the formation of subcultural groups. While the village is also ethnically diverse, intermarriages are common. Ethnicity and kinship offer people some flexibility in self-identification, rather than reinforcing subcultural divisions.

Seppälä found that people in Kilimahewa were involved in many different income-generating activities (IGAs). Some were extractive activities like brick-making or selling firewood; others made an income from artisan work or crafts, making pottery, mats, or rope. Others provided some kind of service – midwifery, haircutting, shoe repair; or traded in food, clothing or other goods. While many people were turning to IGAs as a supplementary source of income, there were large differences in the returns for different activities. IGAs reinforced the effects of differentiation in access to good quality land, rather than acting as an equalising mechanism.

People combined several IGAs in 'livelihood strategies' – 'accumulation', 'reproduction' and 'growth'. These strategies were structurally similar, in that they all involved diversification ('straddling'), but successful and unsuccessful households differed in terms of what went into the package. Differences in managerial skills also mattered: not everyone had the ability to estimate risks or time their activities well. Lindi District is quite a poor region and, apart from trading, few IGAs in Kilimahewa offered any potential for accumulation and growth.

By 'livelihood strategy', Seppälä refers to:

> an allocative process which includes continuous decision-making on three issues. The first is the social composition of and cohesiveness of the household. The second is the socio-cultural resource base – the allocation of

resources to non-material values. The third is the material resource base – the allocation of resources between agriculture, wage-labouring and income-generating activities. All these 'managerial' issues have a direct impact on the involvement of the household in the informal sector ... while the participation of a person in an IGA depends crucially on the other sources of income that he/she and other household members can generate, it depends equally on the structure and cohesiveness of the household as a social system, and the cultural orientations of its members.

(Seppälä 1996: 569)

The insight that particular livelihoods may depend on the structure and cohesiveness of households is explored in Chapter 4. Seppälä's main concern is to show that different IGAs were associated with subcultures within the village – Islam, Christianity, the poor, youths, members of the CCM party. Competition for access to resources was linked to competition between socio-cultural spheres. The livelihood strategies people followed also reflected their differing cultural orientations. For example, older people were concerned to preserve their accumulated kin-based networks to support a gift economy, while younger people relied more heavily on money. Christians were a stigmatised minority, more likely than others to engage in low-return extractive activities on the village margins. There is an echo here of the two, culturally distinct accumulation strategies Sender and Smith found in Lushoto District, pre-capitalist and capitalist, which competed with each other for resources (Chapter 1, p.25).

Conclusion

The case studies from Zambia, Zimbabwe and Tanzania confirm the points made in the discussion of multiple livelihoods at the beginning of this chapter. The one clear generalisation to be made is about livelihood diversification. In an apparently growing trend, rural households in regions across the continent cannot rely on farming alone to make a living. Increasingly, they cannot depend on receiving a flow of remittances from the urban areas, either. Falls in formal sector employment and rising living costs are squeezing urban dwellers' incomes. Some people who might have thought that they were leaving the rural areas for good when they moved to the cities in the past are returning to the countryside. Many young people try their luck in the cities for a while, but eventually come back empty-handed. Others are too discouraged to look for an urban job at all.

The case studies show the importance of location-specific factors in making it possible for people to construct multiple livelihoods. Ex-migrants and their relatives in Mambweland found a niche, albeit a precarious one, in the complementarities created by the international border. Differentiation between regions around such location-specific opportunities is likely to become increasingly important, particularly if the share of farm incomes in total incomes continues to fall. Not everyone in a region can take advantage of such opportunities, however. It seems that differentiation in access to non-farm incomes is marked in many regions. Sometimes this may stem from longstanding differences between households in their access to education (as the Koguta case study will show). At other times, where there are significant social and cultural divisions in a locality, as in Kilimahewa, these may create barriers to entry into different sources of livelihood (cf. Reardon 1997).

The efforts of individuals and households to construct a portfolio of livelihood sources are shaped by calculations about risk and returns to labour. But these decisions are also mediated by social relations. Households do not make decisions about livelihoods – individuals make these decisions through processes of negotiation, *diktat* or the construction of consensus with other individuals. The efforts of some household members to earn an income may, like Moore and Vaughan's commercial farmers in their attempts to capture their wives' labour, actually reduce other members' ability to diversify their livelihoods. The Chivi case shows that the same people may follow more individualised strategies under some conditions and more co-operative strategies when conditions change.

Intra-household relations have important effects on people's ability to construct livelihoods. They also play an important role in determining whether these livelihoods are sustainable. Different kinds of livelihoods set up different issues for relations between household members. They may be more compatible with some kinds of household relations than others. It is to these issues that we now turn.

Notes

1 Reardon defines non-farm income as income from non-farm wage employment, local non-farm self-employment and migration income. Farm income is income in cash and kind from cropping and livestock husbandry.

2 Of course, many markets in supposedly market-based economies are also embedded. This is this case for markets that evade state regulation and taxation ('moonlighting', many domestic services). It also applies to many sectors in the labour market, where information about job vacancies is transmitted through social networks. Dichotomies between embedded economies and 'pure' market economies overplay the impersonal nature of market institutions.

3 Grown and Sebstad developed their livelihood systems framework to analyse women's poverty, but this framework can readily be adapted to look at household poverty and livelihoods.

4 We return to this point in relation to understanding intra-household relations in the following chapter.

5 See also Cliffe 1978; Turok 1979; Vail 1983, cited in Berry 1993.

6 Pottier points out that Watson's explanation for the resilience of agricultural production shifted within the study from one based on the gender division of labour to one based on descent and residence. Mambwe agriculture did not have a rigid gender division of labour, so women could substitute for the labour of absent men. In the conclusion to his study, Watson shifts his ground. In drawing a contrast between the impoverishment of Bemba agriculture in the face of labour migration and the maintenance of subsistence production and social cohesion in Mambwe territory, he argues that the residential stability promoted by patrilineal descent in Mambwe villages is the key to the contrast. Matrilineal villages, like those of the Bemba, lacked cohesion (Watson 1958: 226–7; Pottier 1988: 5–6).

4 Rural livelihoods and gender

Earlier chapters have shown the upheavals of economic change experienced in the African countryside in recent years and their impact on rural people's efforts to maintain their livelihoods. Most people living in rural areas have faced falling incomes. They have had to deal with the effects of crumbling infrastructure and deep cuts in government spending on health and education. Many people in regions which depend on remittances from migrant labour have seen these tail off as urban incomes fall. At the same time, farmers in other regions have been drawn into growing crops for the world market, often on a contract basis. Changes in the sources of livelihood raise new issues in rural households, and intensify old ones. How do shifts in the ways in which household members make a living affect relations of power and authority within households? How do domestic power relations shape the ways in which resources are used? Who is responsible for finding money for food when remittances dry up? Does this responsibility give the right to decide how the money should be earned and spent? Who should work on the new crops? Who should get the crop income? What should it be spent on?

In order to put together a framework for answering these questions, we need to turn to general models and concepts of household relations, particularly concepts of bargaining. To what extent are the interdependencies on which these concepts rely to be found? Are there, rather, pressures leading to greater female dependence, or towards the fragmentation of the household? Understanding gender relations within households involves linking the dynamics of relations within households to broader structural changes going on at the local, regional and national levels and to the operation of state and market institutions. These processes have economic, legal, political and ideological dimensions.

The new household economics and bargaining models

Theoretical debates about power relations in households provide a starting point for looking at links between gender and livelihoods. A first approach, which is still common among economists, treats the household as a single decision-making unit with a joint welfare function. This leaves decisions about the allocation of resources within the household unanalysed (Becker 1981). Critics of neo-classical models of the household have stressed the importance of contractual relations and bargaining (Folbre 1986, 1994; Sen 1990; Evans 1993; Kabeer 1994; Hart 1995). Sen has argued that household members' access to consumption goods is determined by bargaining in a process of co-operative conflict. The household is formed and sustained so long as both parties have more to gain than they would in their breakdown positions (the positions different members would be in if the household did not exist). In societies where women's bargaining position is weak, as in South Asia, women are systematically discriminated against in the distribution of resources, with seriously detrimental consequences for their welfare. Bargaining power is shaped by what the breakdown position would be. Sen suggests that this is not the only determinant, however. The perceptions different household members hold of what their interests are, their valuation of their own wellbeing and their sense of what are legitimate distributions within the household affect the positions they adopt. All these appear to be related to the acknowledged contributions different members make to the household. In some societies, notably in South Asia, women may devalue their own entitlements and consider quite inequitable distributions to be legitimate. Kabeer has suggested that Sen's insights should be extended to bargaining, conflicts and inequalities between household members in decisions about production, as well as in consumption decisions (Kabeer 1991).

Sen, drawing on Boserup (1970), suggests that there are quite marked cross-cultural differences in the social value placed on women and, therefore, in women's bargaining power within households. Where women's acknowledged contribution to the household economy is greater, particularly where they earn an income, they appear to have greater access to resources. Sen argues that these differences show up in welfare data. In regions where women are more involved in economic activities outside the home, their life expectancy relative to men is generally higher than in regions where female economic activity rates are lower. Do women have greater bargaining power when their contribution to household livelihoods is mediated through markets? Does money talk? Not always. There is also plenty of evidence to suggest that women often find it difficult to translate income-earning into bargaining power. Women's incomes are often appropriated by their husbands, or are used to fulfil what are seen as women's responsibilities to provide for children (Whitehead 1984, 1990).

Bargaining models have the potential to offer some powerful insights into household relations. They force us to think about how the conditions under which people form households affect their power to influence decisions. They suggest that changes in the conditions affecting members' breakdown positions will change their bargaining power. They also imply that changes in what indi-viduals get out of household membership, relative to their breakdown position, may change their incentives to remain within the household. These issues are particularly important in Sub-Saharan Africa today. Deteriorating economic conditions require households to construct livelihoods from a medley of different resources and activities, throwing into question the material basis of household construction and maintenance, as well as power relations within households. Household composition and relationships appear to be in flux across the region. Many men are unable to provide an income to support wives and children. Marital breakdown, and the growing numbers of women who never marry, call into question the existence of a material base for the conjugal unit in some regions.

What bargaining models do not offer is a framework for understanding how individuals' breakdown conditions are formed (and how they change) and how the terms on which they bargain are formed, and change. These models also assume that what is going on inside households can be conceptualised in terms of bargaining, even if in only an implicit form. In order to address these issues, we need to link individuals' strategic behaviour within households to broader processes beyond the household, which set up the rules of the game, and the ideologies which legitimate them. These processes include institutions which define access to and control over household resources (property rights; inheritance practices; norms concerning marriage and divorce; other rights and responsibilities which members recognise towards one another), as well as gender ideologies. Several writers have suggested systematic cross-cultural differences in these broader processes. Sen points towards one possible generalisation in his comparison between regions where the female/male activity rate is high (Africa other than North Africa; East and South East Asia) and regions where it is much lower (West Asia; South Asia; North Africa), arguing that these contrasts are linked to differences in women's entitlements (Sen 1990). Kandiyoti offers a framework for understanding regional contrasts in the concept of the 'patriarchal bargain'. This heuristic device contrasts two different systems of 'set rules and scripts regarding gender relations, to which both genders accommodate and acquiesce, yet which may nonetheless be contested, redefined and renegotiated' (Kandiyoti 1988: 286). In one, 'classic patriarchy', which applies to North Africa, the Middle East, South Asia and East Asia, households are patrilocal and corporate, women are subordinate to men and to their mothers-in-law, and the patrilineage appropriates women's labour, making their production less visible. The prospects women have of greater domestic power when they gain the status of mother-in-law encourage them to accept their subordination and in turn, reproduce the conditions for the subordination of their daughters-in-law. Open conflict is likely to be rare. In the other system, found in Sub-Saharan Africa, men have relatively little responsibility for supporting their wives. Women face greater insecurity, which is matched by areas of autonomy that they try to enlarge. Gender conflict occurs where men threaten this autonomy. This distinction between different forms of patriarchal bargains is very schematic. Kandiyoti is aware of this and offers these models as ideal types only, for expansion and refinement with empirical content.

There are some possible problems with the concept of a patriarchal bargain. The term may conflate processes through which individuals strike implicit bargains with situations where bargaining is explicit. One may be present without the other. The question of how much women are able to follow strategies, and have agency, is still left open. It is unclear how far individuals are able to act strategically, rather than simply read from the 'script'. These objections cannot be addressed at this level of abstraction. Gender relations are material, social, ideological and moral. They knit together divisions of labour, sexuality, affection, ideas about rights and responsibilities, ideologies about what men and women are and how they should treat one another. Kandiyoti has more recently explored these issues, looking at the assumptions about consciousness, personhood and subjectivity on which models of bargaining rest (Kandiyoti 1998).

One way of understanding how these different dimensions of gender link up is to make comparisons between household relations in different regions. Kandiyoti's contrast between 'classic patriarchy' and Sub-Saharan Africa is an interesting starting point, but it is pitched at a high level of generalisation. Contrasts between regions are partly a matter of differences in opportunities for constructing livelihoods and partly the effects of regional differences in resource control and decision making.[1] Gender relations may vary greatly between households within a single locality. Households with different bases of livelihood show different kinds of relations between men and women. These relations may not necessarily be characterised by explicit bargaining. They may, in many cases, be better conceptualised in other ways. Changing opportunities for constructing a livelihood may alter the terms on which men and women attempt to get access to land, labour and cash income, shifting household relations between interdependence and dependence. They may also increase pressures for households to fragment, or to reconfigure around relationships other than marriage.

Gender relations are also shaped by legal changes in rights to productive resources and by the practices of state and market institutions. The cases show the impact of individualised land tenure on women's authority in farm decision making.[2] They also demonstrate the negative effects for women of the tendency for contracting institutions to deal with male household heads. Case studies from migrant labour economies and former migrant labour source areas reveal the profound impact of changes in labour markets on household relations and household composition.

The cases which follow represent a spectrum of different livelihoods, based upon the extent to which households rely on farm income or income from wages and trading. They range from quite prosperous and highly commercialised farming households, to households engaged in less commercialised farming, and then on to households in migrant labour economies and in regions experiencing the effects of retrenchment in urban labour markets.

Commercial farming areas

Agricultural commercialisation creates interdependencies. Men take control of land, production decisions and the income from crop sales. All this makes women more financially dependent on their husbands. They want access to crop income: they need cash to provision the household. Where the surplus income is substantial, they have an interest in its being spent in ways that benefit the household as a whole. This gives them a strong incentive to co-operate in production for the market. It also means that their incentives to co-operate will be much weaker if men take all the crop income, or spend it on their own priorities for consumption or investment. When this happens, they have to find cash in other ways. On the other hand, men need women's labour power. Many pure cash crops (such as tea, coffee and tobacco) are very labour intensive. If they are neglected, yields fall substantially.

These points of interdependence can also define the focus of conflict. Where a household is growing a pure cash crop, production for the market competes with food production. If the crop can also be consumed (as in the case of maize), then decisions also have to be made about the proportion of the crop to be sold. Interdependencies may encourage husbands and wives to co-operate, or they may intensify conflict. There is often friction over the allocation of the woman's labour time and the ways in which crop income is spent. These issues may be resolved in quite different ways, as can be seen in the following case studies from Kenya and Zimbabwe. Women may withdraw their labour from commercial production; they may have strong incentives to co-operate; or they may have little choice but to defer to their husbands.

The first case, from Kericho District in Kenya's Rift Valley Province, shows how women in households growing tea have become more reliant on money from their husbands. It also shows the tensions that can arise over the allocation of women's labour and the impact these can have on production. Both partners have some bargaining power, but women's ability to withdraw their labour from the commercial crop and work only on food crops is a weapon of last resort. They are able to deploy it because they can argue that their obligation to grow food must come first.

Contract tea farming in the Rift Valley Province, Kenya

Farmers in Kericho District, Rift Valley Province, grow tea on contract with the Kenya Tea Development Authority (KTDA). When Dorthe von Bülow and Anne Sørensen did their research in Kericho in the 1980s, they found that 30 per cent of the tea plots in their sample were being neglected because of labour problems. The reasons for this neglect could be traced to conflict over men's access to women's labour time and women's access to the crop income. The KTDA assumed that men were the farmers and made crop payments over to them, so that they controlled the crop income. In fact, tea production depended crucially on women's labour. In households where there was not enough money to hire labour, women felt pressured. They had also lost access to labour from beyond the household. Before tea was introduced, women used to exchange labour with neighbours and friends through labour groups (*morik*). Because tea demands so much time, women neglected *morik* in favour of working on their own farms. They could not then call on *morik* to help with food production. Pressure of time in farm production had other effects. For women, maize growing took priority over tea. Making farm work the priority forced them to take time away from domestic work and also from their own cash-earning activities and made them more and more financially dependent on their husbands.[3] It also made the issue of how much women should work on the pure cash crop the focus of domestic conflict.

Conflicts over women's labour time explained low output in tea outgrower schemes. Sørensen and von Bülow (1990) found that women tried to meet their obligation to work on the tea plot if their husbands acted as responsible heads of households, spending the tea income in the common interest of the household. Underproductive growers turned out to be farmers with labour conflicts brought on by conflict over the use of the crop income.

Women did not openly challenge men's control over the income from tea. Both men and women tended to subscribe to a gender ideology legitimating male control. This stood in contrast to earlier ideas about women's rights and responsibilities in provisioning their houses. But men's need for their wives' labour in tea production created interdependencies that gave women some bargaining power (von Bülow 1992). Women did have some potential to bargain with their husbands over labour and access to income, because, unless the husband could afford to hire in labour, the wife was indispensable. Most women in tea-growing households thought that they did benefit from tea production under male control and withdrew their labour only because of pressure of work or where their husbands' neglect was severe. In other words, women accepted the gender division of power, and strategised within it.

Men could not necessarily translate their control over crop income into undisputed authority over the allocation of household labour. Women did have some ability to control the allocation of their labour between the pure cash crop and crops which can be consumed. Because of their responsibility to provide food for their households, they could argue for the need to take labour away from the pure cash crop when there was pressure on labour time. Tension was closely linked to the use of income from the pure cash crop. The issue of how surplus from the production of the pure cash crop was used seems crucial in determining women's willingness to make a large input into the crop. Where the wife could be persuaded that the surplus would benefit the household as a whole, she made the contribution (cf. Odaga 1990 pp. 177–8). Men's position as landowners put women in a weak position to make claims to receive, control or allocate income from contract crops, but this was partly mitigated by their ability to negotiate their labour input. Where women perceived that they were not benefiting from the surplus, they withdrew their labour and retreated into food production. Women's potential ability to withdraw their labour had a double-edged quality, however. It could lose them any chance of access to a surplus. Nevertheless, the threat to do this did confer some bargaining power over the use of the surplus.

In Donna Pankhurst's study of maize growers in Zimbabwe, women in commercial farming households seemed much less able to bargain. The crop could be both sold and consumed, so the crucial issues concerned the proportion of the crop sold and women's access to the crop income. Women's labour was not so crucial, because commercial farming households were headed by migrant men, who could employ labour. These factors reinforced the effects of changes in marriage which had reduced women's autonomy and given them little domestic bargaining power. Women in poorer households had strong incentives to remove labour from household production and, where possible, invest resources outside the household.

Commercial maize farming in Mashonaland East Province, Zimbabwe

Pankhurst carried out research in Murasi,[4] a village lying about 70 kilometres from Harare in one of Zimbabwe's most successful smallholder farming areas, where the main food and commercial crop is hybrid maize. Farming offered the main way to earn an income locally, but many men were migrants elsewhere in the country. There were striking inequalities between households in Murasi. Some people were quite prosperous, with 'Western, urbanised' houses and a range of consumer goods. At the other extreme, some households had almost nothing in the way of possessions and their children showed signs of under-nutrition. Gender relations contributed to these inequalities.

There were several reasons for inequalities between households. Land was scarce and quite small differences in landholding could have a significant impact. Access to off-farm income and household composition were similarly important. Better-off households received remittances or had an adult male present. Without an adult male, households suffered labour shortages. Households without remittances could not compensate for the absence of adult men by hiring in labour. At one extreme were households with migrants who sent regular remittances, employing labour and selling crops. At the other were poor households, often made up of women and children, who could not grow enough crops to feed themselves and who worked on the farms of the better-off.

While women living in different kinds of household experienced quite different constraints, they all faced a fundamental insecurity, since divorce could leave any of them destitute. Because the husband–wife relation had become more important than other links, they could also expect little support from outside the household. It had become socially unacceptable to complain publicly about one's husband. Women in all types of household found it extremely difficult to control resources in the face of opposition from their husbands.

The consequences of divorce for women are confirmed in Jennifer Adams' study of female agricultural workers in Masvingo Province, southern Zimbabwe (Adams 1991). Adams found that *de jure* female-headed households had significantly fewer productive resources and were more likely to be doing casual farm labouring. Single women had few alternatives to this kind of work. Married women had a greater chance of access to land, cattle and other resources, through their husbands.

The focus of conflict and negotiation between husbands and wives differed between poorer and better-off households in Murasi. There was less overt conflict between husbands and wives in better-off households than in poorer households. In better-off households, women needed money to pay for consumption goods, and to pay for labour inputs for the farm. Securing access to their husbands' remittances was crucial and they needed to persuade their husbands to make a commitment to invest in the farm. They had a strong stake in avoiding conflict. This was intensified by the prospect of the consequences of divorce. Women faced losing their children, their access to land, and their possessions. These considerations gave women little room for manoeuvre. Some women agreed to let the man control the farm product. More often, they made an elaborate show of deference and drew attention to the efforts they were making on the farm. While a woman might then control the organisation of production and in some cases control the product, the boundaries of this autonomy were set by the husband, who could check up on her activities when he visited. Women in this group did not think that they were comfortable, materially or financially. They blamed their husbands, saying that they were unreasonable about spending their earnings on household members or about controlling the product of their wives' labour. Women tried to overcome their difficulties by working hard on the farm, by trying to get more resources from their husbands, and by earning their own incomes. Sometimes they did this secretly.

These responses reveal women's weak bargaining position. Unlike tea growers in Kericho, male landholders in Murasi who were intent on farming commercially were mainly employed migrants. They could overcome labour shortages through using remittances to pay for labour. This reduced their dependence on their wives. They could also supervise their wives fairly easily, because the village was reasonably close to the urban labour market. Women's potential for bargaining was also circumscribed by the marriage contract. Women were expected to provide labour demanded by their husbands. Unlike women working on a pure cash crop, they could not easily argue that their obligations to grow food precluded their work in commercial farming. The distinction between commercial and food production followed the allocation of labour, rather than preceding it. This put them in a structurally weak position, because they could not afford to bargain too hard for access to crop income. The consequences of divorce were too severe. In terms of models of household relations, men's superior market position, coupled with husbands' and wives' markedly different threat points, left women with few alternatives to ostentatious acquiescence. This is an interesting finding. It has long been clear that off-farm income is important for rural households' ability to invest in commercial farming. Multiple livelihoods are the norm, rather than the exception. Comparing commercial farming in Kericho and Murasi suggests that the very ability of households to engage in multiple livelihoods is double-edged. If only men have this ability, their command of cash income may greatly increase women's dependence on their husbands. But if women can also earn an off-farm income, then they may have more bargaining power. However, this greater power may, too, be double-edged. The case study from Dixie, Mpumalanga (p.94) shows that female-headed households need to be able either to find a source of income compatible with childcare responsibilities, or, if the woman is a migrant, they need to have a reliable source of childcare. In the absence of any of these possibilities, households were highly unstable and children suffered severe neglect.

The experience of people in poorer households in Murasi shows that men may contest women's ability to get involved in off-farm activities. Indeed it was poorer households, particularly where the husband was present, which experienced the most conflict, sometimes violent. In households which did not grow enough crops to feed themselves, it was necessary for women and men to sell their labour to better-off households. For women, this forced them to neglect work in the home or on the farm. Because husbands often opposed their doing this, it was a risky strategy. Open defiance could be grounds for divorce. So they often worked in secrecy, or in return for food and clothing. Husbands expected wives to use their earnings to finance household consumption, rather than to make investments, presumably because investments were seen as a male domain. Women often subscribed to these priorities, but they sometimes also diverted resources away from the household for safe-keeping. This diversion would impoverish the household still more. So women's inability to make investments within their marital households encouraged them to remove resources from the household. Gender relations deepened their poverty.

Commercial farming in Central Kenya: dependence and challenge

Jean Davison and Fiona Mackenzie carried out studies of gender and farming in Central Kenya. Davison compares gender, farming and land in Mutira Location, Kirinyaga District, in Central Province with Chwele Location, in Western Province (Davison 1988; Mackenzie 1993). Mackenzie's study was carried out in Murang'a District in Central Province.

Davison looked at the impact of changes in land tenure on women's access to land and on women's production. The Swynnerton Plan undermined women's relative self-sufficiency, increasing their dependence on men. This change worked through three processes. The granting of individual land ownership rights to male household heads marginalised women's usufruct rights to land. Furthermore, because land was needed as collateral, women found it difficult to get agricultural credit. Finally, the capitalisation of agriculture and the turn to export crops encouraged by Swynnerton further marginalised women's labour in food production. Food crop production has continued to be identified culturally with women, while commercial crop production is seen as a family affair, often orchestrated by the male household head. In both Chwele and Mutira, men's involvement in agricultural labour depended on the scale on which the crop was grown and the extent of its commercialisation. Household divisions of labour varied, but there were no cases where women were the sole producers of commercial crops and none where men were the sole producers of food crops. Both land and women's labour are scarce resources. Women's work on commercial crops takes them away from food production and women in both areas spend more time on commercial crops than food crops. In a situation of growing land scarcity, more land has been converted to commercial crops. These developments undermine women's ability to fulfil their household's food needs and make women more dependent on income from crop sales and wages. However, it is men who dominate access to the income brought by commercial crops. In the majority of cases, men control income from production. Men play an increasingly dominant role in the management of property, control of land and the distribution of goods and services. In other words, because men have been increasingly conceptualised as the owners of land, they can successfully lay claim to deciding its use and to the income derived from it. Women's labour input alone is not enough to give them enforceable claims to crop income, because their husbands can claim that the land is their property. State policy has reinforced these trends, with marketing boards directing payments to land owners.

Women have challenged this development. In 1983, women in Mutira protested collectively about the payment of cash bonuses by the KTDA to (male) landowners, rather than to individual harvesters. They were successful and the policy was changed. In Murang'a District, Central Province, where coffee is the dominant commercial crop, women responded to the same problem by withdrawing labour from coffee production. In Murang'a, closer to Nairobi than Kirinyaga District, male out-migration reaches over 60 per cent of male household heads in some locations, leaving many women with sole responsibility for coffee production, but with restricted access to coffee income. The District Farmers' Cooperative Union makes crop payments to shareholders, the large majority of whom are men. There was widespread dissatisfaction among women with these arrangements. Payments from husbands to wives were unreliable and not commensurate with their labour input. Many wives preferred to work as daily wage labourers on nearby coffee estates, where they could have greater control over their labour power (and, presumably, over their incomes). The quality of smallholder coffee declined. In an attempt to raise productivity, the Union decided to encourage members to open joint accounts.

These cases from Central Kenya show how important land ownership rights are for legitimising men's power over labour allocation and control over cash incomes from farming. Women's actual contribution to commercial production, in terms of their labour input, does not alone give them enough power (or the right kind of power, ideologically legitimated power) to get access to income commensurate with the labour they put in. There does not seem to be much bargaining going on over this in these cases. Where challenges have come, women in Mutira went over the heads of their husbands to the KTDA and women in Murang'a opted to withdraw their labour. State and co-operative institutions, concerned with maintaining agricultural output, then intervened to modify the balance of power in household resource allocation.

So while commercial production creates interdependencies between husbands and wives, these do not in themselves lead to bargaining. They may just make it difficult for men to maximise production while also keeping control of the crop income. The absence of bargaining may lower output. One important issue, therefore, concerns the imperative household members perceive to co-operate in commercial production, rather than withdraw or reduce their contribution to it.

Suzette Heald's study of tobacco farmers in two areas of Western Kenya shows how the structure of authority in households, and gender divisions of labour, can affect the responses people make to the demands made by an intensively grown commercial crop. As in Kericho and Murang'a, many Teso women withdrew their labour from the commercial crop when they felt that they were not getting the benefit of the cash income it brought their husbands. In Kuria households, on the other hand, there was a greater sense of common purpose and more flexibility about responding to the heavy demands made by tobacco growing. Heald also suggests that Teso women had more incentives to co-operate when growing flue-cured tobacco, because the rewards were greater. Households may become more unified when it is clear that all members can benefit from co-operation.

Contract tobacco farming in Western Kenya

Teso farmers in Busia District, Western Province, and Kuria farmers, in South Nyanza, began growing tobacco on contract with British American Tobacco in the mid 1970s. For the first few years, their earnings were high, but poor harvests in 1984–85 and 1985–86 cut these earnings. Many farmers in Teso reacted by switching to other crops, but this did not happen among the Kuria. This difference in response relates to contrasts between the two areas in terms of household composition, authority relations and divisions of labour.

Teso households are smaller than Kuria households. The household development cycle is short, with households dividing when sons marry. Larger domestic groups probably gave way to conjugal units with the shift from livestock to agriculture after colonial incorporation. Cattle holdings had sustained the authority of family heads, which was undermined. The Kuria, in contrast, maintained their herds, together with an extended patrilocal domestic organisation, in which domestic units are typically two-generational. Teso households are characterised by sharp divisions of interest between husbands and wives, and in domestic responsibilities. There is conflict over men's access to women's labour, while men control the farm income that this labour generates. The intense demands of tobacco growing bring these conflicts to the surface. Women are under pressure to work on a crop which directly competes with their food crop production, when they often have good reason to fear that the income from tobacco may not be made available to buy food. So when tobacco yields fell, these conflicts led many households to drop out of the scheme.

This response was more common among households producing less remunerative fire-cured tobacco. In households producing the higher-earning flue-cured crop, there seems to have been a shift towards a more unitary household economy. The much larger earnings associated with flue-cured tobacco gave husbands and wives a strong incentive to co-operate. Household members may move towards a more co-operative relationship when the material reward for doing so is greater and where women perceive that they have a reasonable prospect of benefiting. Just as changes in what individuals get out of household membership may affect their incentives to form households, so changes in the rewards for co-operation may restructure household relations. There is a clear parallel in my own research in Kisumu District, where differences in the corporateness of domestic units are closely related to household differentiation around incomes and bases of livelihood.

As well as being larger, Kuria domestic units have a strong ideology of joint interest. The homestead's herd of livestock provides a common interest in symbolic and real capital and bolsters the authority of the head. Domestic units also display much more flexibility over divisions of labour in agriculture, much of this being organised collectively. Homestead heads' authority is considered to rest on consent – it is hegemonic, in contrast with despotic relations in Teso households. Homestead heads have the authority to ensure continued production of tobacco when yields fall, while greater household size and flexibility make tobacco less competitive with food production. The connections between household relations and livelihoods are far from simple. Contract tobacco farming rests on quite different household forms and divisions of authority and responsibility in the two regions. In Teso, it exacerbates conflict over control of labour and access to cash incomes in households where the rewards are not high enough to put a premium on co-operation. Among the Kuria, homestead members have an already existing sense of common interest, which has a material and ideological basis. This greater corporateness helps to maintain the authority of the homestead head in organising production, and also allows for flexibility in the face of the labour demands of tobacco growing. In Teso, women's work in tobacco production competes directly with their responsibility for growing food. At the same time, the rewards from tobacco production raise tensions over the use of surplus.

Where households are intensively involved in commercial production, access to women's farm labour and the uses of crop income are important issues. Where income from selling crops is small, the dynamics of household relations are quite different and other issues become the focus of concern. In regions where labour migration is common, gaining access to migrants' incomes takes on central importance in people's strategies for putting together a livelihood.

Migrancy and farming in Chivi communal area, Zimbabwe

The work of Scoones and colleagues in Chivi communal, Southern Zimbabwe, was discussed in Chapter 3. Cecile Jackson carried out research in two areas within Chivi, Madangombe and Gwendomba, between 1988 and 1989 (Jackson nd.; 1997). While both areas are prone to drought and quite densely populated, Jackson chose Madangombe and Gwendomba as examples of greater and lesser environmental stress, respectively. In both areas, livelihoods are based on farming maize and small grains, on livestock products and on remittances from migrants. In 1988–89, about a third of Madangombe women were in marriages with an absent husband, against 20 per cent in Gwendomba. The local economy in Chivi was under strain. Unlike Murasi, Chivi did not take part in the boom in smallholder commercial farming that followed Zimbabwe's independence. Money was scarce and it was difficult to find off-farm work (Corbett 1994; Scoones *et al.* 1996).

Jackson was concerned to understand conjugal contracts in the two areas and the ways in which they were changing. Looking at the long term, it was difficult to construct an accurate account of marriage in the region before the colonial period. Later changes in the social relations surrounding marriage were so contentious that any account of earlier periods given by informants is likely to be coloured by contemporary conflicts. Jackson argues that, immediately before the colonial period, the process of marriage appears to have been managed by the wider kin groups involved, giving women little room for manoeuvre. As in so many other migrant labour economies, the control of elders was undermined by the ability of young men to earn money for bridewealth. Marriage became much more of a contract between individuals. Women themselves played an important role in this change. They did this through opportunistic use of marriage laws introduced in the colonial period. Later, they resisted marriage to old men by marrying young migrants who could pay bridewealth and by their own attempts to move away from the rural areas.[5]

By the late 1980s, men needed to be employed in order to marry. This was not because of the need to find money for bridewealth. Indeed, many men had never actually paid much bridewealth. They went through the motions of negotiating, but then did not pay. Women colluded in this in order to make sure that resources coming into the marital household stayed within it, rather than being passed on to parents. Women seemed to have made employment a criterion of marriageability for men.

The withering away of bridewealth made marriage less stable. Women had less to lose, because they were in a stronger position to retain rights over children. This gave them more agency within marriage and made them readier to end unhappy marriages. It also gave them more access to migrants' incomes. So the commoditisation of male labour, by eroding bridewealth, increased women's room for manoeuvre within marriage. These processes had gone farther in Madangombe. Its greater environmental degradation was a concomitant of the much more intensive commoditisation of life in the area. Divorce was more common there and it seems that women were initiating divorce more often than in the past. The reasons they gave reflected a growing concern with their rights to respect and consideration. As well as more frequent divorce, recently formed conjugal units were splitting off from larger domestic groups earlier in the household developmental cycle, sometimes as early as a year or so after marriage. Young women were becoming increasingly unwilling to accede to the demands of in-laws for their labour. They were also capturing their husbands' remittances more effectively. Men used to send their remittances to their parents, but now wives had gained the right to receive and control them. Jackson argues that women had played a major role in pushing for these changes. The growing nucleation of domestic groups both benefited women and had been stimulated by them. Conjugal contracts were not only not static, but women had some agency in changing them to their advantage.

In a later account, Jackson stresses that women in Madangombe had to manoeuvre carefully. They could use their stronger bargaining position to put pressure on their husbands to pay for labour-saving goods and services, such as scotchcarts for carrying fuelwood and mechanical grain milling. However, they had to avoid accusations of laziness – a common ground for divorce when initiated by men.

Jackson carried out her research in Chivi before the drought of the early 1990s. As we saw in Chapter 3, Scoones and his colleagues found that larger households in Chivi were more easily able to diversify and were therefore more secure than smaller households. So the tendency towards earlier splitting-off of younger households which Jackson observed may in fact increase vulnerability in times of drought. The increased individualisation in Madangombe may also not be sustainable during droughts. During the 1991–92 drought, people who had followed more private strategies for generating and accumulating income in more plentiful times combined their efforts with others, within the household and beyond. The growing scarcity of formal sector employment may mean that households also become more vulnerable in times when rainfall is adequate. All this suggests that greater individualisation is double-edged. Women may benefit from release from the demands of their in-laws, but this also may weaken their ability to adapt to environmental stress.

Migrant labour and the decline of migrancy

Where migrancy is in decline, new points of strain, conflict or negotiation emerge. The focus of domestic conflict lies around questions of who contributes to the household's livelihood, and how much. Questions about the authority, rights and responsibilities these contributions confer are contentious. The answers can reflect relations of dependence, interdependence or household fragmentation. The generalisation that household relations are characterised by bargaining is too broad to capture the dynamics.

Migrancy and retrenchment in Lesotho

For the last century, life in Lesotho has been dominated by labour migration to the South African mines. Under pressure from shortages of labour and other resources for agriculture, Lesotho moved from being a prosperous, commercialised farming economy to an impoverished labour reserve. Murray in the 1970s and Ferguson in the 1980s studied the heavy dependence of Lesotho households on mine labour incomes (Murray 1981; Ferguson 1992). Murray and Ferguson looked at the conflicts that arose between migrant husbands and their wives in Lesotho over the uses to which mine wages were put. Husbands and wives had different priorities. Male migrants were concerned to make sure that their incomes were used for long-term investments that could be used to support them and their households on their retirement ('building the house'). They assumed that women could finance day-to-day household reproduction from farming and occasional, small-scale income generating activities. They were opposed to their wages being used to pay for short-term needs. Because they were working so far from home, however, men found it difficult to control the ways in which women used cash.

Ferguson showed that men dealt with this problem by investing in cattle. Far from clinging irrationally to a 'bovine mystique', Lesotho men were following a rational investment strategy. It was generally accepted that cattle should not be sold off merely to meet day-to-day needs for money. The barrier between cash and cattle could be crossed readily only in one direction – from cash to cattle. So men's investments in cattle were relatively insulated from demands from the rest of the household for cash. They were a retirement fund. At the same time, men resisted the idea of their wives' earning money outside the household. According to the prevailing gender ideology, 'Sesotho tradition' dictated that a husband should provide for his wife. Men should control cash (Sweetman 1995). This ideology arose in response to the centrality of migrant incomes to household economies in Lesotho. It is similar to prevailing views about men's responsibilities in Koguta (Chapter 8).

More recently, families in Lesotho have had to deal with the impact of retrenchment in the mining sector. The numbers of Basotho mine workers in South Africa peaked in 1987, at 126,000. Since then their numbers have fallen substantially, reflecting the growing tendency for the Chamber of Mines to recruit workers from within South Africa (Crush 1995). There has also been a more general reduction in demand for mine labour. Caroline Sweetman and Caroline Wright carried out studies of gender relations in the context of this retrenchment (Sweetman 1995; Wright 1993).

Sweetman looked at the impact of retrenchment on gender relations in the households of ex-miners. She found that ideologies of gender were resistant to people's changed circumstances. Before retrenchment, the dominant gender ideology constructed the household as a unit. People's behaviour often conformed to this model. Women who did not have access to land for farming had time on their hands that they could have used to earn an income. Instead, many conformed to a 'patriarchal bargain' in which wives got cash from their husbands, rather than earning a separate income. The activities which women did undertake (such as beer brewing) were ideologically constructed as an extension to 'reproductive' subsistence agriculture. Sweetman found that people often held on to this patriarchal bargain after retrenchment, even though the economic rationale for it had disappeared. Ex-miners' wives were more likely to be unemployed than to migrate for domestic work, for example. Where they did earn an income, they were restricted to low-paying informal sector activities.

But Sweetman also found signs that this ideological resistance sometimes masked tacit acceptance of changes to domestic divisions of labour. Women negotiated a delicate balance between increasing their contribution to the household's income without challenging prevailing gender norms. This balancing act made it possible for women to take more responsibility for the household economy, while systematically devaluing their contributions. If the income-earning activities women took up were ideologically constructed as 'female', the money they brought in would also be ideologically constructed as limited to subsistence functions (rather than contributing to 'building the house'). This devalued it. The tendency for women's contributions to house-hold income to be devalued has been found in other contexts. Working-class women's wages in Britain are often either constructed as 'pin money' or are ear-marked for spending on children (Whitehead 1984). There were other changes. When men were in work, they tended to make decisions about how the wage was used. Now joint decision making was becoming more common. Women were taking on more managerial involvement in household finances. However, Sweetman stresses that it would be misleading to assume that this change was wholly positive for women. There are several reasons for caution. Joint decision making did not necessarily imply greater equality. Women might devalue their own welfare, while there might be considerable conflict over priorities. Women's greater responsibilities might well amount only to 'stretching' the households' limited finances to meet subsistence needs. Equally important:

> As income goes down, the residue which remains may be viewed as more valuable than before and this, together with time to kill and the need for self-assertion, may lead to male input into previously 'female' gendered areas of responsibility.
>
> (Sweetman 1995: 36)

Men had clung to the ideology of male headship and, ironically, were now much more able to make it a reality, because they were around to enforce it. It was too early to tell whether this state of affairs could persist in the face of the fact that many Basotho women were making a much greater financial contribu-tion to their households. It is a good illustration of why domestic power relations are not a simple reflection of members' contributions to household finances.

Wright's research shows the strains this divergence between material contributions and domestic power relations can set up. Women were the main contributors to household finances in 38 per cent of her rural sample and 35 per cent of her urban sample. Yet men tended to monopolise available income, even when their wives had generated it. There were also strains because men and women had quite different priorities. Women wanted to spend more of the household's income in the household domain (on food, clothing, education, housebuilding), while men wanted to spend it in the public domain, such as drinking. They often demonstrated that they were still head of the family through violence. All this opens the question of whether marriage remains women's best option.

Wright does not want to overstate this. Non-marriage may be an outcome, more than a choice. Growing numbers of Basotho women were not marrying, but this can also be explained by a demographic deficit of young men and women's superior educational attainments. An explanation in terms of the undermining of the material basis of the household is plausible, however. It would suggest that the strains being experienced by men and women in Basotho households are similar to the processes undermining households in other parts of southern Africa.

Poverty and household instability in South Africa

South African research on household poverty emphasises how material insecurity can undermine people's ability to sustain domestic relations. Research carried out in QwaQwa in the 1980s explored in detail the impact of relocation on the formation and composition of households and the stresses that often led households to fragment and reconfigure. John Sharp shows how a division opened up within QwaQwa between households with a reliable source of income (usually from wage employment in common South Africa) and those without (Sharp 1994). In the 1980s, the distinction between workers with stable employment and the marginalised majority hardened, partly due to state policy and partly due to the changing labour needs of employers. In the latter group (by far the larger), the relationship between men and women was 'turned on its head'. Previously, men in regular employment had taken on a breadwinner role, with gender relations governed by an ideology of female domesticity. This is commonly found in African households whose livelihood depends on circulatory labour migration. Just as Sweetman found in Lesotho, women with unemployed husbands took on more and more responsibility for finding money and it was activities open to women that became crucial. Many took up shebeening and street trading. Others were able to find work in the factories that were set up in QwaQwa at this time, mainly offering work to women. In Lesotho, women took on greater responsibility for household reproduction without challenging prevailing gender norms. In QwaQwa, these changes were much more explicit and men contested them. In 1984, unemployed men rioted and attacked women factory workers, chasing them away and protesting that the jobs should be given to men (Bank 1994). However, Leslie Bank argues that it is not enough to argue that this violence simply resulted from the undermining of old, patriarchal social relations. Drawing on an argument set out by Henrietta Moore, he makes a case for looking at the links between gender, violence and identity. He suggests that violence takes place when men are undergoing a crisis of masculine identity. Men resort to violence to reaffirm their identity. In the peri-urban slums of QwaQwa, that identity had been bound up with the ability of men to mobilise female labour for their own income-earning projects. Women's decisions to work outside the home challenged this ability. Moore argues that the QwaQwa case shows a crisis in expectations surrounding the role of father and provider. Marriage is becoming an increasingly unattractive option for both men and women. Men delay marriage until they feel they can support a family. Masculinity is redefined and played out in street gangs, rather than through taking on the role of provider. Women's growing reluctance to marry reflects their seeming inability to control resources within marriage. Marriage undermines their security (Moore 1994). The material basis of marriage is being undermined, or, in the terminology of household economics, the gains offered by marriage are falling below people's threat points.

This does not necessarily mean that households fragment, or that social relations fall away, but the form they take may change. Isak Niehaus' research in QwaQwa focused on the role of kinship in household formation in Phuthaditjhaba, the only town in QwaQwa (Niehaus 1994). Niehaus challenges the assumption that the erosion of marriage inevitably leads to the formation of matrifocal households. He found, instead, that sibling relations had become the key principle underlying the formation of households. There seemed to be material reasons for this development. Many women and men in Phuthaditjhaba were commuting long distances to work in the town of Harrismith, while we have also seen that many women worked in local factories. Long-distance migrancy, more often by men, was common. Niehaus confirms Bank's findings about marital conflict over women's employment and men's inability to act as providers. He found a high rate of divorce and growing reluctance to marry. A significant number of households were formed around sibling relationships. Gender relations seemed more flexible in these households, making it easier for men to take on childcare for their working sisters. Sibling relationships, being more informal and flexible than marital relations, were more compatible with the demands of wage labour than the roles of husband and wife. Nuclear families were often worse off than extended family-based households. They had higher dependency ratios and often dispersed their dependants to other households. In other words, they could not maintain themselves independently of wider networks.

In QwaQwa, sibling-based households could be sustained by access to waged work or trading income for men and women. In the remote settlement of Dixie in Gazankulu (now Mpumalanga) in the Eastern Transvaal, options were much more restricted and the consequences for household relations were profound.

In this area, the only significant employers are nearby game reserves. Competition for these jobs is intense, and this is carried over into conflict between individuals and families within Dixie. Men compete for work on the game farms; women compete for access to men's wages through sexual liaisons. This competition puts strain on domestic life and makes marriages fragile. Households and household membership are unstable. All this threatens women's ability to support their children. Neglected wives may leave and return to their parents, or they may be 'chased away'. Even then, a woman's parental household may be unwilling to provide much support, unless she herself can earn a wage. In order to do so, she will have to leave her children in the care of relatives or older children, and these carers may neglect them. Children's insecurity is reinforced by their 'social invisibility'. Adults tend to give priority to their own needs (Kotzé 1992). They are also unreliable, because their own positions are so precarious. For many children, deprivation and insecurity are the norm.

J.C. Kotzé's research in Dixie concentrated on children's responses to this extreme insecurity. Because they cannot count on any one adult to provide food and shelter, they have to find these for themselves. Toddlers are particularly vulnerable, but once children pass the age of four or five they develop long- and short-term strategies, building networks to replace their dependence on mothers and other relatives. Children work at other homes in return for cash, food or accommodation. They also forage in the veld. But the world of children in Dixie is not a Hobbesian nightmare. Children put a great deal of effort into building social bonds with other children. Kotzé was struck by the way that children immediately share any food and clothing they can get hold of. 'The meaning of something one receives is contained in the way in which it can be put to use in securing and enhancing one's social value' (Kotzé 1992: 158). What matters is to bind other children to oneself, creating obligations that may later translate into material resources. In contrast, children's allegiances to their families tend to be pragmatic and mobile. Their relationships with adults are shaped by high levels of mistrust. Here, then, households are unable to sustain the reproduction of children, who must reproduce themselves through social co-operation.

Kotzé stresses that the reasons why households in Dixie are so unstable, particularly the competition for access to wages from the game reserves, are location-specific. However, it is worth thinking about why these conditions should have such devastating consequences. Kotzé suggests that the crucial factors are men's access to higher wages, coupled with women's ability to find employment (which makes it possible for them to establish homesteads independently of men and to enter relationships with other women's husbands). Women need access to men's wages in order to subsist, but they are free enough from male control to be able to compete for them with other women. So, if women were even more dependent on their husbands, they would not have the option of infidelity and households would perhaps be more stable. In other contexts where men dominate access to wages, women redistribute these wages through selling them commodities. One reason why this does not happen in Dixie may be that there is not much scope for diversifying women's livelihoods. Villagers were resettled in Dixie from the game reserves in the early 1960s. Kotzé states that they have access to grazing land (for stock which are, presumably, owned by men), but he does not mention any agricultural activity. Dixie is also an isolated settlement, remote from markets, so there is not much scope for trading. So women's options are either to compete for men's wages, or to go off to work themselves.

Summary and conclusions

Any general conclusions about household livelihoods and gender drawn from a comparison of these case studies must be tentative. The regions involved are very different from one another, not only in terms of the economic processes at work, but also in terms of the kinds of social and institutional changes they are experiencing. Differences in marriage, in gender ideologies and in property rights are all-important and are not linked to differences in livelihoods in any simple way. Comparisons and generalisations can be made, to the extent that the case studies show the effects of various structural factors. These include livelihoods, and incentives that they may or may not provide for men and women to co-operate and factors which make domestic groups more or less unitary or stable. They also include factors that may make it easier for men to take control in households, or for women to act assertively.

The cases can be contrasted in terms of the livelihood possibilities they offer. If we look across different kinds of livelihood, we see that there are contrasts in terms of the issues they set up for household relations and the pressures they generate. These have implications for the extent to which household members are interdependent, for the relative power of women and men within the household and for household stability. These contrasts are important not only between regions, but also within them, because not all households in a region have the same possibilities for constructing a livelihood.

The first kind of household economy, commercial farming, sets up clear interdependencies. Men need access to women's labour, while women need access to the crop income. These interdependencies may give rise to co-operation, or they may generate conflict. If there is conflict, it can have a variety of outcomes, ranging from the wife's withdrawing from commercial production to her showing elaborate deference to her husband. Household members are more likely to co-operate if there is a clear benefit. In Kenya, this happened in Teso households earning relatively high incomes from flue-cured tobacco. Where that benefit is more doubtful, perhaps because earnings are lower, or where women are excluded from payment procedures, a co-operative outcome is less likely and households may be less unified. Overt conflict over the allocation of women's labour time may occur. Where the household is growing a pure cash crop, as in the Central Province cases from Kenya, it seems somewhat easier for women to withdraw their labour, arguing that they need to ensure a supply of food for the household. The Murasi case from Zimbabwe shows that where households are selling a food crop, it is much more difficult for women to make a case for choosing how to allocate their labour. This case also shows that women have less bargaining power if men can afford to hire in labour to replace them. Changes in marriage that have reduced women's autonomy also seem to be very important in Murasi. The fact that migrant men can supervise their wives relatively easily also strengthens their control.

This last point raises wider issues. African commercial farmers very often rely on off-farm income, especially wages, to finance their investments. This means that many commercial farming households combine wage labour with farming in the ways described for Murasi. From the perspective of male migrants, this combination of activities is feasible only if women's farm labour can be supervised. Men tend to be reluctant to delegate management of commercial farming to their wives. Gender ideologies that stress women's unreliability and their inability to handle money properly reinforce this reluctance. If migrants cannot easily supervise what happens on the farm, they are much more likely to invest their earnings in activities that do not need such close supervision. Children's education and rental housing fit these requirements well. Male migrants from Lesotho deal with the problem by buying cattle to 'build the house'. The combination of migrancy with commercial farming is much more likely to happen in areas easily accessible to labour markets than in more remote regions. We see this factor at work in the contrasts between Central and Western Kenya made in Chapter 5. Migrant men also have to deliver on their part of the bargain. Because women farmers in Murang'a were dissatisfied with the coffee income their husbands handed over, they preferred to work as wage labourers on coffee estates. So gender relations shape the possibilities for dynamic links between farm and off-farm income. In households and regions where men cannot easily control the farm enterprise, these links are likely to be weak. Access to the migrants' wages becomes the focus of attention.

Access to migrants' wages can either increase or reduce the domestic power of women. It can also change the composition of domestic groups. In Chivi, Zimbabwe, wages have made young men more independent of their elders, and able to avoid paying bridewealth. This makes it easier for recently married couples to split off from larger domestic groups, and for young women to capture their husbands' remittances. In Chivi, migrancy encourages the formation of nuclear households, because co-operation with others can only reduce income, by allowing them to make a claim. In Koguta, women who are dependent on migrants' earnings seem to have less domestic power than women who have to find money for themselves. This may reflect the greater impoverishment of the local economy – in the households receiving regular remittances women's dependence on migrant incomes may be much greater and their bargaining power consequently much smaller.

In regions heavily dependent on migrant labour, the decline in the availability of work for migrants raises tensions over responsibilities for maintaining livelihoods and the powers that these responsibilities should confer. In Lesotho, continuities in gender ideologies that keep men in a provider role co-exist uneasily with women's taking on responsibilities for bringing in money. Changes in domestic responsibilities call into question deeply held assumptions about the roles of husband, father, wife and mother. When changes in responsibilities are explicit, men may try to reassert their masculinity through violence, as in the QwaQwa case studies. Women's contributions are devalued and where there is joint decision making it may only mean that women take on the stress of stretching the household budget. The tendency for women to lose control over their income when it enters the household is widespread. When unemployed men appropriate their wives' incomes, marriage starts to look like a more unattractive option than some of the alternatives. Marriages break down, or women become reluctant to marry in the first place. These outcomes do not represent social breakdown. The South African literature emphasises the growing importance of matrifocal households, as multi-generational households share resources and domestic labour. Niehaus shows that households may also reconfigure around sibling relations, particularly if these are more adaptable to the women's need to combine wage labour with childcare. However, residential instability is also likely to become more common.

The last case study, from Mpumalanga, shows dramatically how there can be clear material reasons for household instability and fragmentation. In this case, women compete for access to men's wages, but also themselves have the option of constructing a household based on their own earnings. Because they have to resort to labour migrancy in order to earn this income, women find it difficult to build households that last and children are seriously neglected by men and women alike. Stable households require the existence of either a dependable income (such as a wage) or the potential for members to combine multiple livelihoods. They also require the presence of members who can provide dependable childcare. The presence of these factors does not guarantee the absence of tension, or even violence, but their absence seems to be disastrous, particularly for children.

Household bargaining is, therefore, only one of a number of possible processes through which men and women construct livelihoods. The interdependencies created by commercial farming may encourage household members to bargain over access to labour and crop income or they may serve only to define points of conflict. Women may not be in a position to bargain, or they may withdraw co-operation. Whether or not household members bargain, co-operate, or even remain committed to the household will be influenced by incentives, in the form of the rewards they expect for co-operating, or the penalties they face for withdrawing co-operation. The outcome also depends on many different factors, including household composition, divisions of labour and property rights, authority relations and gender ideologies.

How these processes were registered, negotiated and argued about in Koguta is a major theme in the second part of the book. We shall see how people in Koguta reacted to the pressures imposed on them by labour migrancy and the ever-growing poverty of the rural economy. Before we hear about the experiences of individuals, however, we need to look at how the major contours of their lives were set. The central element in this story is the stagnation of Luoland in the twentieth century, which made it the impoverished backwater it is today.

Notes

1 Heald's comparison of tobacco farming in Teso and Kuria communities (1991) demonstrates this point.
2 The case studies by Davison (1988) and Mackenzie (1993) deal with this issue explicitly.
3 Von Bülow and Sørensen stress that women's increased labour burden preceded the introduction of tea and so cannot be explained by this change. The introduction of maize as a cash crop, the growth of commercial dairy production and of male wage labour all increased the demands on women's labour time.
4 Murasi is a pseudonym. Pankhurst (1991) and Pankhurst and Jacobs (1988).
5 See Schmidt (1992) and Barnes (1992) for detailed accounts of changes in marriage and gender relations in colonial Zimbabwe.

Plate 1 Picking coffee, Western Province, Kenya

Photograph by Tim Allen

Plate 2 Sorting through the coffee crop, Western Province, Kenya

Photograph by Tim Allen

Plate 3　Drought in Zimbabwe – women collecting weeds for food amongst the ruined maize crop

Photograph by Tim Allen

Plate 4　Preparing the soil for planting, Zimbabwe

Photograph by Tim Allen

Plate 5   Making pots, Zimbabwe
Photograph by Tim Allen

Plate 6   Shopkeeper, Zimbabwe
Photograph by Tim Allen

Plate 7 A Koguta woman being helped to hoe her field by another woman from the same *dala*. *Jodala* who got on well with each other still did this, though the larger *saga* groups exchanging labour had been replaced by groups of people who occasionally hired themselves out for a wage (*Plate 8*).

Photograph by Elizabeth Francis

Plate 8 Hired labourers working the fields

Photograph by Elizabeth Francis

Plate 9 A boy drives his family's herd of cattle home in the late afternoon. Children were expected to work in the fields and around the house after school and at weekends.

Photograph by Elizabeth Francis

Plate 10 A young girl preparing vegetables

Photograph by Elizabeth Francis

Plate 11 Two of the ways in which men made an income locally: thatching (*Plate 11*) and making bricks (*Plate 12*)

Photograph by Elizabeth Francis

Plate 12 Men making bricks

Photograph by Elizabeth Francis

Plate 13 Sondu market: women selling grain. Most of the food sold at Sondu had come from outside Kisumu District.

Photograph by Elizabeth Francis

Plate 14 Sondu market: a tailor

Photograph by Elizabeth Francis

2

Poverty and livelihoods in Western Kenya

Figure 5.1 Western Kenya

5 Koguta and Western Kenya

A journey from Nairobi to Kisumu

Koguta East Sub-location is in Kisumu District, Nyanza Province, about 150 miles north-west of Nairobi.[1] Kisumu District, along with Siaya, South Nyanza and Migori Districts, encircles the Winam Gulf; the small portion of Lake Victoria that lies in Kenya. Lying at the edge of the belt of Highlands running north-west from Nairobi, these districts are the home area of the Luo people. Much of the literature about smallholder farming in Kenya has focused on Central Province, the centre of Kenya's smallholder boom. Looking at Western Kenya gives a very different perspective on Kenya's 'success story' in smallholder farming.

The most dramatic way to grasp the contrast between the two regions is to take the overnight train from Nairobi to Kisumu. In the early evening the train first goes through the almost Asian landscape of Kiambu: hilly, green and crammed with tiny field plots. Here and there are larger farms and modern, red-tiled houses. These often have tiny, wooden houses for farm labourers nearby. Soon, however, the landscape is erased by darkness. In the morning, the scene is different altogether. The railway line runs through the north of Kisumu District, between the steep escarpment of the Nandi Hills and the hot, dry Kano Plains. This is the Nyanza Sugar Belt and sugar-cane fields surround the stations along the line. The rest of the landscape, especially in the dry season, looks almost desolate. There are only a few tall trees and many of the fields of grain and pulses look neglected. The land is dotted with compounds of mud houses with thatched or iron roofs, all looking very much alike. There are just a few brick or stone houses and people are much more poorly dressed.

Many parts of Western Kenya are lower and drier than the Central Highlands. They stand in a Lake Shore Savannah agro-economic zone that receives only 30–40 inches of rain annually, and sometimes much less, in two rainy seasons (De Wilde 1967). Because the rain often falls in violent thunderstorms, much of it is lost through evaporation, while low-lying areas are often flooded. There are also large areas where the soil is not particularly fertile, brown, friable clay (ibid.). Because of these poor natural endowments, a great deal of the region is unsuitable for the high-value agricultural activities (coffee, tea and grade cattle) that underpin the prosperity of the higher income districts. Apart from Kisii District and a few other higher and wetter areas, most of Western Kenya is too low and too dry to produce these commodities. Where coffee can be grown, distance from the major markets and poor infrastructure make marketing difficult. People trying to sell green vegetables and horticultural crops in the higher areas face the same problems. Kisumu town has been economically stagnant for many years. After the collapse of the East African Community in 1977 its port virtually died. For these reasons, Kisumu is not a major market for farm products. Instead, the main market for food is Nairobi. However, there are not many all-weather roads in the region and transport is slow and difficult, especially in the rainy seasons. Outside Kisii, the main crops grown are maize, beans and pulses. There is a local market for these crops, but returns on produce sold are low.

Far from being exporters of food to other parts of Kenya, many parts of the region regularly experience food deficits. There is a hungry season in the months before the long rains harvest in July and August. Even the most casual observation shows that much of the food in the rural markets of Kisumu and Siaya Districts comes from outside. Sondu market, four miles from Koguta, is a nodal point for the import of food into Nyakach from Kericho and Kisii Districts. Outside Kisumu town, almost all the people with permanent off-farm employment are teachers or public servants. There is little rural manufacturing and most trading is very small scale and unremunerative.

Since the beginning of colonial rule in Kenya, the history of Luoland has been about labour migration. Luoland has been a source area for migrant labour for the plantations, farms and towns of the rest of Kenya since before the First World War and the region's economy has been dominated by the impact of migrancy on farming and rural livelihoods. Luoland is also one of the poorest areas of Kenya. These two great facts of life in Luoland – migration and poverty – are more complicated than they appear on the surface. Migrancy did not have to bring about the decline of farming. Some migrants did very well on the urban labour market; some parts of the region could have supported the high-value cash crops being grown in Central Province or nearby Kisii. Commercial agriculture failed to take off for a host of reasons. Some concerned the region's political economy. Luoland was far from the major markets for crops and livestock in Central Kenya. Colonial officials wrote off Luoland as a migrant labour reserve and then it was marginalised in national politics after Independence. There was also a growing shortage of land. But agriculture also fell into decline because people had other priorities for their money; because educating one's children looked like a safer bet for the future than planting coffee bushes and because migrant men preferred not to entrust their wives with management of a commercial farming operation when they could visit only two or three times a year. All this makes it important to understand how decisions about livelihoods at the local level and within rural households were linked to what was going on in the regional political economy. Processes of change at these different levels, the household, the locality and the region, have been intricately related and need to be examined in turn.

Migration and economic decline in Luoland

Until the 1930s, the colonial economy of Kenya was dominated by the acute problems of labour supply facing the Government and the European farming sector.[2] The basic problem for the Government and settlers was that they could not attract enough labour voluntarily at the price they were able to pay. African societies gave most people access to the means of production. The growing markets for livestock and crops made it relatively easy for African farmers to meet their need for money. The wages offered were extremely low, since the Government and settlers were short of cash. The Government was supposed to be self-financing. Settler farming in Kenya was inefficient and used a great deal of capital. Settler farmers were chronically in debt. They paid low wages and working conditions were usually poor (Clayton and Savage 1974; Van Zwanenberg 1975; Stichter 1982).

For all these reasons, most Africans preferred to earn money by selling crops and livestock. The Government responded by extracting labour coercively. Hut and poll taxes forced African men to earn money and labour 'recruiters' were sent into the African reserves. In Nyanza, this began around 1900, once 'pacification' was completed. Recruitment was easier in Nyanza than some other regions, because there were several chiefs in the region who were powerful enough to be able to recruit labour on the Government's behalf. This was one major reason why Nyanza became a centre for the supply of migrant labour (Stichter 1982: 5–19).

Not all labour was coerced. In Kowe, in the north of what is now Kisumu District, 'considerable numbers of men voluntarily sought employment outside their home areas from the earliest days of British administration' (Hay 1972: 165). Some migrants were the first members of a new elite of mission-educated young men who left to find work as clerks and teachers. Whole households also moved to European farms around Muhoroni, in Kisumu District. Moving in search of grazing land, they entered into labour tenancy arrangements. Other voluntary migrants were young men who wanted money to buy consumer goods and cattle for bridewealth. They often worked for just a few months and returned when they had earned enough money to buy the goods that they wanted.

After marriage, however, many of them began to migrate for much longer periods, some working for as many as thirty or forty years. Ondiek stopped working as a migrant in the 1930s, but Obiero worked at various jobs in the Rift Valley until the 1950s. By the 1930s, it was no longer necessary to use coercion to recruit labour, however, and both Nyanza and Central Province had become large-scale suppliers of labour. In Koguta, the majority of men seem to have been involved in the labour market by then.

Obiero (born 1906) went to work as a sisal cutter near Mombasa in the late 1920s because he wanted to buy good clothes. He stayed for two years and then he came back, but his brother took his clothes, so he went back to Mombasa for another year to get some more.[3]

In the First World War, Ondiek (born 1902) did various contracts for six months at a time. His aim was to buy a belt, a hat and money that he gave to his father to buy cattle. 'Then you would come back and dance, and return when the things were worn out.'

It is very likely that migration by young men before the 1930s, whether forced or voluntary, did not have much impact on the local economy. Young men did not play much of a role in farming. Apart from cattle raiding and hunting, their main occupation was herding and children could replace them in this work. The move to mass migrancy by adult men was much more significant, because it meant that their labour was lost to farming. It probably happened because the expansion of commercial farming in Nyanza was so limited, quite unlike the expansion of commercial farming seen in Central Province in the 1930s. Returns to resources put into farming were simply lower than returns to other activities. In Central Province, Kikuyu smallholders were enthusiastic adopters of wattle trees (mimosa), whose resin was used in the tanning industry. Forestal, a multinational firm, promoted wattle production in Central Province in an attempt to undercut production in Latin America. Wattle gave high yields per unit of land, yet was not labour intensive, making it an attractive crop in a local economy where land and labour were scarce (Cowen 1975). No such crop was made available to Luo farmers.

Many of the young men who first left Koguta to get money to pay for taxes, clothes and bridewealth in the 1920s continued to migrate for work in the 1930s. Once the need for money was entrenched and it had become clear that farming was not going to provide an income large enough to cover this need, the options narrowed. In the 1930s and 1940s, a few young men were able to stay in the area and prosper, but they were the last to be able to do so. They owed their success more to trading than to farming. They were not selling local grain, either. They took millet from the highlands of Kisii and the Rift Valley to areas of Luoland where many producers had switched to maize. Nobody in Koguta seems to have sold crops on a large scale during the 1930s, although there was also no evidence of significant food deficits once the locust famine of 1931–32 had passed. The picture is one of stagnation, rather than crisis.

However, this famine, together with low commodity prices brought about by the Depression, confirmed to many that wage labour was a more secure means of earning money than farming. The early 1930s seem to have been the crucial years in which many parts of Nyanza became migrant labour reserves. They were also the years in which the loss of male labour from agriculture began to be registered in the production process. Maize replaced sorghum as the staple food. It can be double-cropped and its labour demands are less peaked. Crops with high yields in calories relative to the labour time needed to grow them, like cassava, became popular. The local economy appears to have been quite resilient in the 1930s. This was probably because women increased the amount of time they spent working on the farm (Hay 1972).

At the same time, the need for cash – for clothing, for school fees, for bridewealth – was growing. New items of 'discretionary' consumption were gradually transformed into 'necessary' requirements (cf. Arrighi 1970). This process was self-perpetuating. In order to make sure that they had a cash income in future, households needed to invest any spare money they had in their children's education, rather than in farming. School fees were still quite low in the 1930s, but parents had to buy uniforms, textbooks and stationery. They also lost most of the children's labour in herding and farming. The loss of men's labour power from farming led to a downward spiral of agricultural decline and ever-increasing migration.

Employment and wages

What sort of work did these migrants find? Some found work in the Nyanza Sugar Belt or on the tea plantations in Kericho District, but most went further afield. The major centres of employment were outside the region. In the colonial period, many went to work on the European farms in the Rift Valley. Others went to Nairobi and Mombasa, or to the sisal plantations at the coast. Reflecting the continued lack of local employment, labour migration from Western Kenya is still predominantly long distance.

Until the Second World War, most Africans were employed in low-paid work in agriculture and domestic service. Collier and Lal provide the following figures for 1927:

Table 5.1 African males in the wage labour force (%)

Agriculture	48
Railways	10
Government departments	8
Domestic service	14
Other	20
Total	100

Source: Collier and Lal 1986: 32

From the Second World War onwards, the development of secondary industries increased the demand for skilled labour. The employed labour force grew rapidly during the War and became increasingly urbanised. Employment in private agriculture fluctuated in the 1950s and fell during the 1960s, with the withdrawal of European farmers. It remains a significant employer of labour, however. In 1985, 186,000 workers out of a total employed labour force of 1,174,400 were working in agriculture (Kenya 1986). Wages in this sector are markedly lower than in private industry and the public sector. Private, non-agricultural employment rose in the 1950s, faltered around the time of Independence and then rose rather slowly in the 1960s. The largest increase in employment took place in the public sector (Collier and Lal 1986: 69).

In the 1950s, middle-level positions tended to be given to Asians. Africans' upward mobility was restricted by their inadequate education and training and by racial discrimination in hiring and pay (Stichter 1982: 109–14). However, wage differentials had widened during the War as demand for skilled labour rose (ibid.: 114–17). These higher income groups also became more stabilised in employment and many brought their wives to live with them in town. Many polygynous migrants from Luoland arranged matters so that their wives alternated between living in town and farming in the rural home (Parkin 1978). Unskilled and casually employed workers in agriculture and in non-agricultural employment continued to receive low wages, however.[4] These trends were reversed from the mid 1950s. Partly to undercut the grievances of the growing trade union movement and partly in order to encourage the urban labour force to stabilise, minimum wages were raised and urban real wages increased substantially. By 1959, public sector money wages were 88 per cent higher and private sector wages were 73 per cent higher than in 1949 (Collier and Lal 1986: Chapter 2, *passim*). Real wages continued to rise after Independence, until 1972. 'Africanisation' of middle and upper white-collar jobs and the huge growth of public sector employment after Independence also provided opportunities for a small number of people to earn relatively high wages.

In the 1970s, another transformation took place in the urban economy. Unemployment rose rapidly, a trend which has continued. Many young migrants cannot obtain formal sector work and, instead, are self-employed or are employed in the informal sector. By 1990, at least 250,000–300,000 school leavers were entering the labour market each year, competing for about 100,000 job vacancies.[5] While there is a fairly wide spread of incomes from informal sector activities, many workers receive low, and precarious, incomes. Faced by such discouraging prospects in the urban economy, most unemployed school leavers in the rural areas appear to choose to stay at home and try to make a living locally (Närman 1995).

Migration and farming

Food production[6]

From the early 1930s, many parts of Nyanza were set on a course of economic change that eventually marginalised agriculture as a source of income. The agricultural economy withstood the loss of the labour power of young men in the 1920s, but, from the 1930s onwards, the loss of the labour power of adult men undermined production. As more and more labour was withdrawn from agriculture, people became caught in a downward spiral of agricultural decline and ever-increasing migration.

Other factors contributed to this decline. The fixed boundaries placed on the reserves led to population pressure on the land. Fallowing periods became shorter and declining soil fertility lowered yields. The switch to maize acceler-ated soil exhaustion, because it depletes fertility more rapidly than sorghum. Population pressure also led to the subdivision of holdings. In the 1948 Population Census, Nyakach had a population density of 261 people per square mile. By the 1962 Census, this had risen to 361 (Ominde 1968).

These changes eroded marketed production of crops. But rural households were becoming more enmeshed in the cash economy. They had new tastes – for clothing, tea, sugar and paraffin. They had to pay to send their children to school. By the 1960s, children needed an expensive secondary education if they were to have any chance of finding a permanent job. As more children received at least primary education, most adolescent labour power was withdrawn from agriculture (children often started school at eight or nine years and repeated classes, so many were in their teens by the time they reached the eighth, and last year of primary school).

Gradually, women also spent less of their time farming and more time earning money in other ways. Some households were able to fill the gap between their cash needs and their farm production through remittances, but many turned to off-farm income-earning activities like small-scale trading and farm labouring. All this took women away from their working on their farms, which reduced their crop output even further. Falling real wages in the urban economy and rising underemployment from the early 1970s onwards increased this group of impoverished households.

Migrants in better-paid work might send enough money home to make up for shortfalls in food production, but they made little investment in farming. They concentrated their attention on their children's education. When they bought land, it was outside Koguta. The reasons for this behaviour are complex. As will be discussed in later chapters, they lie partly in the cumulative effects of the decline into a migrant labour reserve (a small local market; poor infrastructure), partly in long-term patterns of accumulation and upward mobility and partly in the dynamics of changing gender relationships.

Statistical coverage of food production in Central Nyanza, though slightly better from the end of the Second World War onwards than in earlier periods, is too patchy to give absolute confirmation of these trends. But there was a steady decline in marketed food surpluses from the end of the War, and the District became a net importer of food (Kitching 1980: 130–2). The first survey that allows us to assess the status of food production in Kisumu District is the Nyanza Province Rural Household Survey, which was carried out in 1970/71 and part of which is reproduced in Table 5.2. It made clear the poor status of food production in Kisumu District relative even to other districts in Nyanza.

Table 5.2 Selected statistics from 1970/71 Nyanza Province Rural Household Survey

(KShs per household)	Kisumu	Kisii	S Nyanza	Siaya
Value of food and sugar production	680.7	160.6	1290.9	1220.7
Value of food and sugar purchased (sold) for consumption	146.0	(-33.2)	(-51.6)	(-36.5)
Costs of farm production of which:	85.6	250.7	141.1	93.9
Hired labour	49.8	118.4	75.2	75.8
Purchased inputs	28.8	88.1	55.2	11.2

Source: Kenya, Republic of (1977) *Nyanza Province Rural Household Survey 1970–71*, Tables 4.2, 4.3. and 4.7

Because of the low value of food production, many households in Kisumu District had to buy a relatively large proportion of their food, hence the high value of net purchases of food and sugar. Another indicator of the relative backwardness of agriculture in the district is given by the figures for the costs of farm production. Evidence that, as well as Kisumu District's weak comparative position, there was an absolute decline in food production comes from a comparison of Whisson's description of the household economy in Central Nyanza in the early 1960s with the current food supply situation in Koguta. Whisson suggested that:

> The prudent mother would expect to have in her home enough food for herself and her children until the next harvest, with possibly a little left over for sale in the market, and a small amount for cash in emergencies.
>
> (Whisson 1964: 104)

Already, then, many households were not producing a food surplus, but, if Whisson's account is more than a normative description, most households were still largely self-sufficient in food. If this was so, there was a marked decline in food production in the next three decades. By the late 1980s, most households in Koguta did not grow enough food to feed themselves for the whole year, sometimes not for more than a couple of months. The food supply was even more precarious in the lake shore areas, where the short rains often fail.

Commercial crops

Nor was there a significant expansion of cash crop production outside the Sugar Cane belt and the irrigated rice schemes in the Kano Plains. Most areas in Western Kenya were largely untouched by the expansion of African agriculture inaugurated by the Swynnerton Plan. They were not ecologically suitable for the high-value cash crops – such as coffee, tea and pyrethrum – upon which the expansion was based (Swynnerton 1954). Production continued to be dominated by grain and pulses. However, because of its elevation, the Nyabondo Plateau was suitable for coffee growing and oilseed production. The first farmers to take on coffee growing began cultivation in the late 1950s. But they were always a very small minority. Few of them took production beyond the very small scale. Oilseed cultivation, while more widely spread, was also in decline in the 1980s. People said that coffee needed too much labour; prices were too low; payments came too slowly and the factory was too far away.[7]

Government research and development efforts have concentrated on cash crops that are not suitable for many parts of Nyanza. Added to this have been serious regional imbalances in the development of infrastructure.[8] These stem from priorities at the centre that both reflect and sustain the political and economic marginality of these areas. Luo politicians were prominent in the Nationalist movement before Independence, but they were afterwards soon marginalised. In a political system so largely structured by patron–client relationships, this political marginalisation of Luo areas has had direct effects on government expenditure in the region.[9]

Off-farm activities

Patterns of off-farm activities, and the kinds of income people could expect to earn from them, have also changed several times. Until the 1960s, trading activities in the region on a scale larger than petty trading were almost entirely dominated by Asians (Fearn 1961). One of the few lucrative trading activities open to Africans in the early years of the period covered was the cattle trade. In Kowe, around sixty miles north-west of Koguta, full-time cattle traders began to operate between 1910 and 1920. Like the early migrants from Kowe, they were among the first Christians. They travelled as far as South Kavirondo (now South Nyanza) and brought cattle back through areas of Central Kavirondo (Kisumu District) where prices were higher (Hay 1972: 177). Some men in the region were able to accumulate by taking advantage of the growing cattle deficit in the local economy and became substantial cattle importers (Cohen and Odhiambo 1989: 77). Cohen and Odhiambo suggest that 'the movement and exchange of surplus cattle … offered owners of herds, transporters, and African brokers a ready entry into the monetized economy of twentieth-century western Kenya' (ibid.). Several Koguta men were involved in the cattle trade and were able to use the profits to make other investments. However, none of them operated on the scale described by Cohen and Odhiambo.

From the 1930s onwards, some men and women from Koguta became involved in long-distance grain-trading. They took maize and millet from Kisii to Kano, a dry area where people were often short of grain. The women who did this traded only occasionally, but several men became regular traders, transporting the grain with donkeys. The fact that some men were trading full time shows that food deficits had appeared in some Luo areas by the 1930s. By the 1980s, with food deficits the norm in Nyakach, grain was bought from Kipsigis traders in Sondu. Women sold it on the spot and in periodic markets in Koguta. Men took the grain down to the drier lake-shore areas and, in the hungry season, around Koguta, still using donkeys. But the trade brought very small profits and aspiring accumulators had since moved on to other activities.

Brick making was a profitable activity from the middle decades of the colonial period because of the growing demand for brick built shops, schools and houses. Several men took advantage of suitable local soil and successfully made brick making their major source of off-farm income. As retailing became more feasible for Africans, however, this and passenger transport took over as the major activities for would-be rural accumulators. Very few of them were successful. Low spending power, competition, lack of working capital and the demands of relatives kept profits low (Fearn 1961; Marris and Somerset 1971). Building real estate became a popular alternative for people without the capital needed to set up a retail business or buy a passenger vehicle. Once the building was constructed, supervision was minimal and rental returns were regular and predictable. These attributes also made real estate a popular choice for urban migrants wanting to invest in the rural area.

Koguta and the Nyabondo Plateau

The journey to Koguta

Koguta East Sub-location, known locally as Koguta, lies on the Nyabondo Plateau in the southernmost part of Kisumu District, standing at the intersection of Kisumu, Kisii, South Nyanza and Kericho Districts. Rising abruptly from the Kano Plains that adjoin the lake, the Plateau is an extension of the Western Highlands jutting out into Luoland.

In colonial Kenya, the African reserves were divided into Provinces, Districts and Locations (which were later divided into sub-locations) and this administrative structure is still largely intact. The British placed boundaries within the African reserves according to demarcations between lineages as they stood at the time of conquest, freezing what was probably a more complex and shifting series of relationships than appeared to administrators, missionaries and early anthropologists (Ogot 1967; Cohen and Odhiambo 1989). Nyabondo was one of the last areas to be settled by Luo people before the British set the boundaries of the African reserves on ethnic lines (Ogot 1967). After fierce fighting, the Kipsigis (the dominant community in neighbouring Kericho District) were driven from the Plateau at around the time that the railhead reached Kisumu (in 1902). Tensions between Luo living on the Nyabondo Plateau and neighbouring Kipsigis flared up again in 1991.

The Plateau stands at 5000 feet above sea level, making it cooler, wetter and greener than the Kano Plains. This is one of the reasons why both the Catholic and the African Inland Missions established themselves on the Plateau in the 1920s. The long rains fall roughly between March and June and the short rains fall roughly in October and November.[10] The Plateau covers approximately 25 square miles and is divided between three sub-locations, Kadiang'a, Kajimbo and Koguta. Koguta covers an area of approximately 8.6 square miles and had a population of about eleven thousand people in 1989. Compared with many other parts of the three Luo districts, Nyabondo had a good range of services. The large market at Sondu was four miles away and there was a frequent bus service to the larger regional markets at Kisumu and Ahero. There were several clinics on the Plateau and a hospital at the Catholic Mission, and, because Nyabondo was a centre for early missionary activity in Nyanza, there was an unusually large number of primary and secondary schools. There was electricity at Sondu and at the Catholic Mission, and electricity poles were being set up on the rest of the Plateau in the late 1980s. There was also a network of stand-pipes, but, more often than not, these were out of action.[11] Dry season water supply was a problem in several parts of the Plateau.

In the late 1980s, most people travelling the 30 miles from Kisumu to Nyabondo squashed into a *matatu* at the bus station. Police roadblocks permit-ting, the *matatu* took the tar road running south east through the Kano Plains and winding uphill to Sondu. If you were lucky, or very patient, you might find another *matatu* at Sondu market to take you up on to the Nyabondo Plateau, but you usually had to walk the last five miles. Most people going from Sondu to Koguta walked anyway, because they could not afford the *matatu* fare. When you reached Nyamaroka marketplace on the edge of Koguta, you would first be struck by the air of neglect. The shops were nearly all locked up, even in the middle of the day; there was no water in the standpipe; and only the teashop and the flour-mill seemed to be doing any business. Depending on the season and the time of day, people would be working in the fields; tending cattle and goats; or sitting outside their houses, making baskets, decorticating sisal, or just talking. Most of the houses were small and square, with mud walls and a roof of thatch or rusting iron. Koguta looked like a place where everyone lived in much the same way and where everyone was poor.

But if you continued along the road that runs through Koguta towards Kokumu marketplace, you would see a cluster of houses that seemed very different from the rest. They had brick walls; some had glass windows and tiled roofs. They would not look out of place in a Nairobi housing estate. You might then have expected to find clear differences between the people living in these houses and the rest. If you stayed longer, however, you would start to realise that the differences between this group of people and the rest were not very great. Most people were farming on a small scale, buying very few inputs for the farm and consuming most or all of their production. Yields were low (three to four sacks for an acre of maize interplanted with beans would be typical) and almost nobody was self-sufficient in food. Every year, most people faced a hungry season when they had to find cash to buy food. Most had migrant husbands or migrant children. In almost all households the major outgoings were on food and household goods. People with school-age children spent a large proportion of their incomes on school fees.[12]

The similarities and the differences; the brick houses and the closed shops, all suggest a complicated story. In fact, there are several stories to tell. One describes the creation of a class of urban professionals from the vantage point of a sub-location on the economic margins of Kenya. Another story deals with the pressures that led most men to spend at least some, and usually most, of their economically active lives living and working away from home. Yet another relates the ways in which many people living in Koguta have been pushed into deepening poverty, while others have been better able to withstand these pressures.

Lineages and settlements

Luo people speak a Nilotic language, Dholuo, and they are believed to form the southernmost component of a group of closely related Nilotic peoples also comprising the Padhola, Acholi, Lango, Alur and Palwo in Uganda and the Shilluk, Dinka, Nuer and Anuak in Southern Sudan. These peoples are believed to have had a common origin in the Bahr-el-Ghazal region of Southern Sudan. Through a long series of migrations, settlers are believed first to have arrived in Nyanza from what is now Eastern Uganda between 1490 and 1600. They gradually spread east and south, across the lake, through expansion, conquest and assimilation (Ogot 1967). In the course of migration from hotter, drier regions in Sudan and Uganda, they shifted from being predominantly pastoralists to a greater involvement in agriculture, particularly after some groups moved into higher land beyond the lake shore (Ogot 1967; Butterman 1979).

Luos belong to virilocal patrilineages. These are considered to be components of larger groupings, which are themselves segments of still larger groupings, culminating in the Luo as a whole.[13] Although assimilation of non-Luo and the complexities of pre-colonial migrations give the lie to this myth, in theory, every Luo has an identifiable place within the segmentary lineage structure. Appendix B provides a detailed account of Luo segmentary lineages.

Koguta contains two closely related lineages: JoKoguta and JoRamogi, which are both major segments of the maximal lineage of JoNyakach.[14] Almost all men living in Koguta consider themselves to be JoKoguta or JoRamogi, but there are also JoKoguta in Koguta West Sub-location, ten miles away beside the Lake at Sango. Many people came to the Plateau from Sango, some moving as lineages, others moving as compounds, or even single houses, but some came from South Nyanza, while others came from as far away as Seme, 15 miles north-west of Kisumu.[15] The sub-location straddles the eastern side of the Nyabondo Plateau and the slopes leading up to it from the Kano Plains. The higher, Plateau land is mostly settled by JoKoguta, while JoRamogi live along the edge of the Plateau and on the slopes below.[16]

JoKoguta and JoRamogi are composed of a number of smaller, closely related lineages. *Jokakwaro* vary in size because of uneven demographic growth and segmentation and a *jokakwaro* can contain anything from two compounds to ten or more.[17] For example, descendants of a particular man, let us call him Oyugi, may have continued to call themselves JoKoyugi, and considered themselves members of a *jokakwaro* bearing that name, for several generations. Eventually, however, with population growth (or sometimes because of disputes), the group became so large that smaller subdivisions, such as those descending from Oyugi's different wives, or sons, became important enough to warrant use of the term *jokakwaro*, while links between different subdivisions of JoKoyugi weakened. Because of differential population growth, lineages would be of differing sizes and genealogical depth. Therefore, major subdivisions between lineages within JoKoguta do not reflect genealogical relationships in a simple or uncontested way and, although informants' accounts of the major subdivisions coincided, their accounts of the placing of these subdivisions within the genealogy of JoKoguta conflicted. On a day-to-day basis, Koguta people worked with a schematic account of genealogical relationships between the larger subdivisions.

In this account, JoKoguta subdivides as follows: the first division made is between JoMari and JoKachungo. This division reflects the large size of JoKoguta *in toto*. For instance, most Luo sub-locations have a social welfare organisation in Nairobi, but JoMari, because of its size, has its own organisation. When interviewed, the chairman of the organisation argued that a single organisation for all JoKoguta would be unwieldy. JoMari are divided into JoKomoro, JoKadero, JoKobala, JoKamuga and JoKojwang. These divisions largely correspond to settlement and landholding patterns within Koguta, with members of the same lineage at this level of segmentation holding adjacent strips of land. The ideal type presented in the literature suggests that the settlement pattern reproduces on the ground the segmentary organisation of lineages. At lower levels of segmentation, landholding patterns usually, but do not always, follow lineage group subdivisions, because of accidents of settlement and the uneven availability of farm land. For example, JoKomoro subdivides into the major lineages (*dhoot*, pl. *dhoudi*) JoKatombo and JoKoyoo. Most members of both hold strips of land adjacent to other members of the same lineage, but some JoKoyoo form a separate cluster near JoKatombo land. These patterns reflect the piecemeal manner in which the land was settled and the uneven distribution of well-drained land on the Plateau. There are also instances in which territorially adjacent *jokakwaro*s are not genealogically closely related. One such is Rarieda (lit. 'the line'), in which I lived. This cluster of settlement, with a resident population of 324 in 1987, contained members of several different *jokakwaro*s within JoKoyoo, but other members of the same *jokakwaro*s were living elsewhere. A similar settlement, Keyo (pop. 119), also consisting of JoKomoro, lay adjacent to Rarieda.[18]

Whisson argues that, although there was a wider system of reciprocity, the *dala* was 'the key economic unit in the society within which it was felt that none should starve unless all starved'.[19] A pre-colonial *dala* typically consisted of a male head (the *wuon dala*; lit. 'the owner of the homestead'), his wives, those sons who had not yet formed their own homesteads, and perhaps his widowed mother and younger brothers. Richer men might also have attached to their *mier* a number of servants (*misumba*) and clients with usufruct rights to land (*jodak*) (Whisson 1964). The basic units in a *dala* were *jokamiyo* (lit. 'people of one mother'), who comprised the membership of an *ot* ('house', pl.

udi). A polygynist would be attached to all the *udi* of his wives, but the titular head of each *ot* was the wife, the *wuon ot* (a monogamous male would be *wuon ot* in his one house). The *wuon dala* was a powerful figure, the ultimate authority and focal point of the homestead. He had nominal control over all the cattle and land attached to the homestead and the right of access to the labour of his wives, their children and daughters-in-law. His power was far from absolute, however. Economic relationships between husbands and wives are discussed in detail in Chapter 8, but it is worth noting here that each *jokamiyo* unit was allotted cattle, small stock and at least one field and had a considerable degree of autonomy over management of these resources and food supply in general. Different *udi* often exchanged labour, though senior women had more access to the labour of junior women than vice versa. Each wife was expected to provide a dish of cooked food for her husband every day. Men ate together, while women and children ate separately from men, but also communally. The basic picture of *jokamiyo* within a *dala*, then, is of separate production units which shared resources at the point of production and consumption on a daily basis and which were all under the authority of a senior man, the *wuon dala*.

Changes in authority and material relationships between members of *mier* have altered the significance of the role of the *wuon dala*. These changes are discussed in Chapter 6. Here, it is sufficient to relate that this person (usually a senior man, but sometimes his widow or eldest son) is at least nominally the ultimate authority and focal point in the *dala*. A euphorbia hedge or a fence often surrounds the *dala*. Each *dala* stands in its own land, and each *ot* has at least one field allotted to it. Although *mier* are not grouped in villages, they are often clustered around well-drained land, as several parts of the Plateau are susceptible to flooding. Adjacent *mier* tend to be very closely related genealogically, since sons usually found their own *mier* around the *dala* of their father. Members of a *jokakwaro* live close together and genealogically close *jokakwaro*s (members of higher order lineages; *dhoot*, pl. *dhoudi*) are adjacent to one another, although there are some exceptions. Within the sub-location, at the level of the largest lineage subdivisions, lineage and territory are largely conterminous, although there are some *jokakwaro*s and compounds which arrived after the majority and which are less closely related to others than most. Informants were reluctant to describe these people as land clients (*jodak*) or even by the more circumlocutory term, guests (*wendo*, pl. *welo*).

The local economy

Land

Much of the Nyabondo Plateau is a plain (*siany*) that becomes waterlogged in the rainy seasons. For this reason, *mier* were clustered on better drained land, leaving the rest of the plain as open grazing land. But land was becoming scarce, so some younger households had established their *mier* on the plain, leaving themselves open to the risk of losing crops through flooding. This happened to several households in the long rains of 1988, which were unusually heavy.

Landholding data are likely to overestimate the availability of farmland, because not all land was arable, while some was being used for housing. Nevertheless, a sample survey of 104 households carried out in 1989 gave the following estimates of holding size. The mean holding size was 4.2 acres, with a median of 2.25 acres. When two households with unusually large holdings (82.8 and 41 acres) are removed from the calculation, the mean holding size drops to 3.15 acres (the median size was still 2.25). Sixty-six households had three acres of land or less, while thirty households had less than one acre. Land-holding size was affected by household developmental cycles. Young men who might expect to augment their holdings a little through inheritance headed some of the households in the last category. Others were older households, most of whose land had been subdivided among their sons. But the general trend was for younger households to face a future with very small holdings.

Households, income and poverty

Mean household size in 1989 was 5.4 people. Table 5.3 shows the different forms of household in a random sample of 104 households surveyed.[20]

Table 5.3 Households, by age and status of husband

age of head	husband present	husband migrant	husband deceased	total
less than 40	16	27	6	49
40–60	19	5	12	36
60+	11	0	8	19
Total	46	32	26	104

In 31 per cent of households the husband was not living at home. This is the kind of distribution one would expect to find in a migrant-labour economy. One other finding from this survey was less expected. In a third of households with a husband present, he was aged under forty. This finding reflects the fact that it was becoming difficult to find regular work in the urban economy. Most of these younger men had spent a few years outside Koguta, often working in the urban informal sector, but had given up and were now back home. Some were trying to make a living from selling crops, or from doing artisan work locally. Others did not seem to have any occupation. Most people in Koguta were poor. The incomes of most households were low when compared to other smallholder regions in Kenya. Tables 5.4 and 5.5 show income and expenditure data collected in a household budget survey in 1988–89.[21] The data show considerable inequalities in income. Households in the upper tercile had incomes three times those of middle-income households.[22] In Table 5.4, house-hold income is divided into five categories: (a) the value of own production of agricultural commodities, livestock and crops; (b) local employment; (c) other rural non-farm activities; (d) remittances and pensions; (e) other sources. The table shows that agriculture was not a very important source of income, accounting for just a fifth of household income. Only three out of nineteen households reported farm income as providing more than a third of total household income. Remittances and pensions accounted for a third of all household incomes. Differentiation revolved around access to remittances and pensions.

Farming

There are two growing seasons, roughly from February to July and from October to December. Most households grew a rather limited range of crops: chiefly maize, sorghum, beans, peas, sweet potatoes and green vegetables and sold little or none of their output. A few households farmed more intensively and sold vegetables, pulses, oilseeds and sometimes maize. Households which invested more time and inputs into farming were all headed by men who were returned migrants.

Simon Onyiero worked as a tractor driver in Muhoroni, in the Nyanza Sugar Belt until he was sacked in 1985, at the age of 35. He decided to return to Koguta while he looked for work and he built a new compound in 1986. He and his wife, Helen, began intensive, small-scale cash cropping. They sold cassava and sweet potatoes and, later, tomatoes and onions. They grew all these and some crops for their own food, on two acres of land. In a good year, they could expect to get KShs.600 from the tomatoes and KShs.420 from the onions. Onyiero also made bricks once a year. He had five cows, three in *riembo*.[23] Onyiero and Helen sold milk from the two cows they were looking after themselves. Onyiero took all the decisions about the farm. Recently, he decided that Helen should start trading in a small way, selling goods like soap locally and at Sondu on all three market days. This would markedly reduce the amount of time she spent on the farm.

Onyiero's decision shows that Helen and he could not get by with cash cropping and brick making. Income from this kind of source came only after long intervals and they needed a regular income. There were some more successful commercial farmers, but even they were operating on a small scale. They were older than Onyiero and Helen and were getting some money from their children. This could tide them over until they sold some crops, and could help them to buy inputs or pay labourers. It was hard to make a satisfactory income just from selling crops. Returns were low and risky. The market for vegetables, in particular, was easily glutted.

There were many reasons why most people could not contemplate growing crops for sale. An older, male-headed household, where the husband and wife

Table 5.4 Average net monthly household income

Name	Farm (cash and kind)		Employment		Other RNFA		Remittances		Other		Total
	A	%	A	%	A	%	A	%	A	%	
Owinys	1394	32	0	0	2879	65	129	3	2	–	4404
Olals	1006	54	0	0	363	19	500	27	0	0	1869
Adhiambo	177	14	0	0	272	21	845	65	0	0	1294
Ombura	161	14	7	1	273	24	711	62	0	0	1152
Akinyi	144	19	40	5	476	61	100	13	18	2	778
R. Ogendo	107	15	0	0	415	57	92	13	117	16	731
Odhiambos	153	22	0	0	77	11	467	67	2	–	776
Osaro	315	47	0	0	0	0	350	53	0	0	665
Odindo	111	18	0	0	127	21	380	61	0	0	618
Okonya	80	14	97	17	258	46	117	21	0	0	554
Akech	52	9	0	0	48	9	400	73	50	9	550
Atieno	142	27	47	9	236	45	47	9	52	11	529
Abuyas	55	10	119	23	207	39	132	25	13	2	526
Owete	153	29	0	0	0	0	373	71	0	0	526
Mboya	82	17	10	2	163	35	147	31	67	14	514
M. Ogendo	59	15	23	6	229	59	41	11	28	7	380
Onyieros	16	5	0	0	299	88	26	8	0	0	361
Ogutu	87	44	0	0	28	14	68	34	17	9	200

Notes:

A amount in KShs.

% percentage of total income

RNFA rural non-farm activities

Table 5.5 Average monthly household cash expenditure

Name	Food		Other cons. goods		Farm		Education		Transfers		Other		Total
	A	%	A	%	A	%	A	%	A	%	A	%	
Owinys	1351	41	474	14	110	3	600	18	321	9	415	13	3271
Olals	353	18	237	12	713	36	186	9	202	10	309	15	2000
Adhiambo	555	53	56	5	40	4	202	19	120	11	76	7	1049
Ombura	431	39	143	13	27	2	375	34	70	6	65	6	1111
Akinyi	431	71	37	6	6	–	26	4	72	12	37	6	609
R. Ogendo	128	18	85	12	3	–	284	40	184	26	32	4	716
Odhiambos	356	28	156	12	41	3	23	2	517	41	157	13	1250
Osaro	365	62	85	14	19	3	0	0	90	15	34	6	593
Odindo	305	71	88	20	3	1	7	2	10	2	41	4	437
Okonya	293	67	71	16	4	1	4	1	24	5	41	9	437
Akech	353	44	105	13	27	3	113	14	91	11	108	14	797
Atieno	369	45	123	15	13	2	65	8	65	8	183	22	818
Abuyas	436	44	145	15	8	1	26	3	130	13	239	24	984
Owete	69	17	82	20	67	16	0	0	73	17	127	30	418
Mboya	168	35	24	5	3	1	0	0	30	6	255	53	480
M. Ogendo	277	55	33	7	8	2	26	5	80	16	83	16	507
Onyieros	250	55	93	20	3	1	9	2	40	8	60	13	455
Ogutu	80	60	12	9	5	4	0	0	9	7	27	20	133

were losing their strength, lacked the labour power needed to intensify the farming operation. They usually would not have the cash needed to buy inputs. The wife of a migrant might also not have enough money to hire labour. Most migrant men were neither willing nor able to send their wives money for anything beyond subsistence needs and school fees.

Most households did not have enough land to grow the crops they needed to feed themselves. Land shortage brought about by population pressure had been intensified by the practice of subdividing holdings among sons at inheritance. Everyone already depended on markets to meet some of their consumption needs. Making a commitment to commercial production would increase that dependence and most people preferred not to put themselves in that position. The opportunities open to them were too poor to make it an attractive option. For all these reasons, people mostly confined themselves to growing maize, beans, cassava and vegetables. Many households sold very small amounts of vegetables and pulses, but very few sold grain.[24] Selling grain was seen as foolhardy; a threat to food security. Use of purchased farm inputs was mostly limited to hiring casual labour and plough teams (very few households had their own). Ploughing was often late, because of the heavy demand for plough teams. However, many households could not afford to hire a plough team. Instead, they worked their land with hoes.

Migration and remittances

People have left Koguta to look for work elsewhere in Kenya since the earliest years of the colonial period. Chapter 6 looks in detail at these migratory movements and their impact on the local economy. Most migrants were men, but many women also migrated; some to join their husbands at their place of work; others going on their own account. Much of this migration was long distance, to Nairobi, Mombasa and towns in the Rift Valley, although some people also travelled a shorter distance to work on the tea estates in Kericho District, or did seasonal work on commercial farms in that District.

Studies of urban migrants in Kenya emphasise the strength of their social and economic ties to their places of origin.[25] It is argued that urban workers maintain close ties with the rural areas because town and country are 'two locations within a single social field' (Ross and Weisner 1977: 370). One of the most important links between rural and urban areas has been remittances by urban workers to relatives in the rural areas. Several studies of remittance behaviour carried out in Kenya in the 1960s and early 1970s found that remittances were quite large. In a study of low- and middle-income urban groups, Johnson and Whitelaw found that 89 per cent were sending remittances and that these remittances were equivalent to 21 per cent of their reported earnings.[26] Rempel and Lobdell found that, of a 1968 sample of recent in-migrants in Nairobi, the equivalent of 13 per cent of the average income of all men in the sample was remitted. Of the total sample, 59 per cent of the men were remitting regularly an average amount of KShs.75. This represented 22 per cent of average income earned (Rempel and Lobdell 1978).

Hoddinott found that migrants from Karateng Sub-location in Kisumu District had a high propensity to remit to their parents, with more than 80 per cent giving money or goods to their parents in 1987/88, but he also found that 54 per cent remitted less than KShs.250 per year, while 66 per cent remitted less than KShs.500 (Hoddinott 1989). Data from the Koguta budget survey were comparable, although the sample was not statistically derived. Three rounds were collected: during the hungry season, just after the harvest and at the time that school fees fall due. Informants were also asked about cash receipts greater than KShs.200 in the intervening periods. Of the eleven households with adult, migrant sons, five received less than KShs.200 in total from all their migrant sons. Remittances from some husbands to wives were higher, around KShs.1000–1200, but three women received less than KShs.300. The number of women receiving low or no remittances from husbands and sons was growing. It is difficult to compare urban and rural-based surveys. The low figures for remittances received in Koguta and Karateng confirm a strong impression gained in Koguta that remittance levels were low. They also seemed to have fallen in the previous decade or so, because of the growing financial difficulties facing urban migrants. People living in Koguta had expectations of support based on their past experience that were not being fulfilled. Parents expected children to support them in their old age; wives expected support from husbands. Falling real wages made it more and more difficult for skilled and white-collar workers to send remittances back to the rural areas. This put a great strain on relationships between urban workers and their rural relatives.

Better-off migrants from Koguta tended to have their families living with them in town. They did not often visit and their material links to the rural area were not very strong. Some did send regular remittances to their parents, but many did not. The reasons for their failure to do so lay partly in the dearth of opportunities for remunerative local investment and in difficulties of supervising a commercial farm or a business operation from a distance. Less well-off migrants tended to have stronger social ties to Koguta. Their wives and children might be living in the rural home, or they might intend eventually to do so. But they found it very difficult to send back remittances to Koguta. Even an annual visit at Christmas, which is the period when many were expected to visit, was becoming prohibitively expensive. Travel was costly. When migrants came home their relatives expected gifts of money, food or clothing. Strikingly few long-distance migrants returned to Koguta while I was living there. Although the region exported much of its labour power to the urban areas, the links between many rural households and urban migrants were so weakened that it would be quite misleading to see them as actors operating in a migrant labour economy.

In Koguta, some young men had given up trying to survive in town and had returned home, hoping to make a living locally. This proved so difficult that some eventually went back to town to try their luck again. They were caught between the collapse of employment in the urban economy and the dearth of opportunities at home.

Other off-farm activities

Grace Akinyi's husband was a carpenter, who retired in 1985, in his late forties. Since then, he had found some small carpentry contracts locally, but he took no interest in the farm. While her husband was a migrant, Grace alternated between living with him, and trading in vegetables, and farming and trading in vegetables in Koguta. He used to send her money monthly. After he retired, she carried on her trading, but ill health had recently forced her to give it up. She now got some money from basket weaving. She also did some farm labouring and occasionally received money from her husband and married daughters – she had no sons. She also sold a few tomatoes, but since her field was less than an acre, there was little scope for cash cropping. Her husband married a second wife in 1981. This woman was still cooking with Grace and working within Grace's field, but she would eventually be given some of Grace's field. The household used to have what she described as an adequate herd of cattle, but two went for bridewealth, some died and others were sold to meet pressing cash needs for food and clothing. The household now had only one cow.

Someone walking through the thrice-weekly market at Sondu and seeing the numbers of traders and the piles of food for sale might first think that the local economy was thriving. Most of the traders in the open-air market were women. Around the marketplace were small shops (*dukas*) and cafes (*hotelis*) and these were owned by men. In the open-air market, the most widely traded commodities were grains, pulses, vegetables and fish. But the bustle in the marketplace was misleading. People were not selling their own produce. Traders bought food from intermediaries who brought it from the surplus producing areas of Kisii and Kericho Districts (or, in the case of fish, from the lakeshore). After the harvest, these sources were supplemented to a small extent by local produce. The food trade was so busy because most households needed to buy food.

Most women did not take part in the food trade and many who did traded only occasionally. Trading was too time consuming. Those who did had no other significant source of cash income and were able to put in the time needed: older women with fewer responsibilities for childcare.

Silpa Osuri traded in maize at Sondu three times a week, making a monthly profit of KShs.600–800. She also raised secondary-school fees by working as a seasonal labourer on a farm in neighbouring Kericho District, going each week in the weeding season and during the harvest. While she did this, she would use hired labour on her own farm. Silpa was able to trade because she had only three children, two of whom had left home. Another woman in the *dala* used to look after her other child.

Rather than trading, most women wove palm frond baskets. The palm fronds (*othith*) came from the coast and the baskets were sold at Sondu market to traders who bulked them up and exported them to other parts of Kenya. Basket weaving could be picked up and put down again as time allowed and the baskets could be sold every week, guaranteeing some regular cash (KSh.20 was a typical weekly profit). Money from selling crops might come in only once or twice a year, while women needed money for food every week. For some women, money from *othith* was a useful supplement to the remittances sent by their husbands. For others, it might be the only source of money, or one among a medley of activities, such as farm labouring, or brewing alcohol.

Poor men might do some farm labouring and small contracts like tree cutting and roof thatching. Those with a little capital might load some maize into baskets and carry them by donkey from house to house, or they might make bricks. If a man had more money to invest (it would almost always be men who had spare money to invest), there were not many options. He could build a house and hope to rent it to a teacher or build some commercial premises in a marketplace and rent them out. These investments were seen as the least risky. If he bought some land to pass on to his sons, it would usually be outside the Koguta, because there was little land for sale locally. He might then move on to open a *duka*, or buy a *matatu*. Few of these enterprises lasted very long. There was too little money floating around in the local economy; and customers wanted to pay on credit. Local businessmen were plagued by short- ages of working capital. Because none of these activities were secure, the investor would need to diversify. But this meant that he would have to find a reliable way to supervise labour.

Jeremiah Oyugi had the best-stocked *duka* in Sondu. Born in 1934, he went up to Form Two at Maseno School, but school fees and travelling expenses were a great problem. After leaving school, he went to the Post Office training school in Nairobi and trained as a telecommunications technician. He did this work until 1959, when he became a broadcasting technician. He later did some City and Guilds examinations and further on-the-job training. He was stationed in Nairobi, Mombasa, Kisumu, Eldoret, Nandi Hills and Kisumu. He was due to retire in February 1989.

Jeremiah had three wives and twenty-seven children. His wives lived with him where he was working for most of the time. He opened his first shop at Ahero, when he was working in Kisumu, from 1966–73. Then he transferred to Nairobi. In Mombasa, in 1975, he opened a shop and his then two wives helped him to run it. In 1980, he opened two shops in Sondu and he opened another one in 1984. Two failed in 1986, 'because of a lack of business ability on the part of the wives who were running them'. The current shop, run by the second wife, was a success, 'because she is a good manager'. He got his initial capital from savings and a bank loan, which had almost been repaid.

He put one of the other wives through commercial college and she was a clerk at a local school. The third wife was at home, 'but I will give her something to do'. When he retired, he intended to start another shop. The profits from the current shop were used to build rental houses. The second wife mostly did this alone, but he also contributed. [27] They had built three houses and a row of rental units, all in Sondu. He had bought plots in Sondu, speculatively, and he had thought about buying a *matatu*. 'People invest in shops, buildings and *matatus* because they are manageable. A large farm involves a lot of work and you have to get a lot of labour, which tends to be unreliable. Many telephone farmers in this area have failed, because their labourers haven't worked properly. You have to spend a lot of time managing a farm, but I don't need to supervise my wife at all.'

Jeremiah had two plots of land, comprising fifteen acres. He inherited a third of both and bought the rest. The wife at home supervised the farm. He sold maize, beans and oilseeds, selling about twenty sacks of maize. He did not grow many tomatoes, 'as that would involve a lot of labour supervision'.

In summary, there was not much to show for JoKoguta's years of labour migrancy. Poverty was widespread, the local economy offered very few opportunities for making a reasonable livelihood and migrancy itself was becoming a less reliable source of money. The following chapter explains why there was so little to show, how long-term changes in JoKoguta's livelihoods undermined farming and left the local economy stagnant. Chapters 7 and 8 explore the stresses of dealing with poverty in the late 1980s.

Notes

1 A sub-location is the smallest administrative unit in Kenya. Sub-locations are grouped into locations. These are part of a division, a district and a province. Koguta Sub-location is part of South Nyakach Location, Nyakach Division, Kisumu District and Nyanza Province.

2 For the early political history of colonial Kenya, see Brett 1973; Clayton and Savage 1974; Van Zwanenberg 1975; Berman 1990; Berman and Lonsdale 1992.

3 All personal names in the Koguta case study are pseudonyms. Lineage names, such as Jokomoro and Jokadero, have not been changed.

4 Stichter suggests that this was one of the reasons for the support given by unskilled Kikuyu workers to Mau Mau (1982: 130–31).

5 Estimates by Närman (1995). The average annual increase in salaried new jobs (including the urban informal sector) from 1984–1990 was 85,000. The figure of 100,000 vacancies includes an estimate for vacancies created by retirements and deaths.

6 In this region there is not a rigid distinction between food and commercial crops. All food crops are potentially saleable. The distinction made here refers to crops that are

mostly produced for household consumption and those, like coffee, where almost all production is sold.

7   Because coffee-processing facilities are far more scarce in Nyanza than in Central Kenya, transport costs for coffee growers are higher.

8   Bigsten documents substantial regional differences in levels of spending on public services. Levels of Government spending were consistently lower in Nyanza than elsewhere. For example, between 1974 and 1978, Government development spending per capita on roads in Nyanza was one fifth of the national average (Bigsten 1978: 384; see also Nyangira 1975).

9   For analysis of post-colonial politics in Kenya, see Schatzberg 1987; Haugerud 1995; Throup and Hornsby 1997.

10 This pattern varies from year to year.

11 The water supply reappeared for one day in 1988 when the President visited Nyakach.

12 Two households participating in a budget survey carried out during fieldwork spent more than 30 per cent of their cash expenditure on school fees (see Table 5.5).

13 The term 'Luo' seems to have come into use only during the colonial period and it was probably a name given to Luo-speakers by outsiders. They were also variously called WaKavirondo, WaPagaya and WaRuguru (Cohen and Odhiambo 1989).

14 The prefix 'Jo-' literally means 'people of … '. In this context it refers to a group united by believed common descent from a single ancestor. *Ka-* (*K-* before vowels) is usually also included in lineage names. Its literal meaning is 'the territory of' a particular person or group, but in this context it means the house of a person. JoKoguta are the people descended from the house of Oguta.

15 The picture of multiple movements (e.g. from Seme, to South Nyanza, to Sango, to Nyabondo) by lineages, compounds and houses is quite different from the ordered movement suggested by Evans-Pritchard 1949.

16 Most of the research was carried out among JoKoguta, rather than JoRamogi, because I was living within the territory of JoKoguta and had much more intensive contact with them. Economic differentiation, one of the foci of the study, was also more marked within JoKoguta. A number of life histories were collected from JoRamogi.

17 Boundaries of a *jokakwaro* are also malleable. Individuals, houses and compounds more distantly related than those mentioned here may sometimes describe themselves as members of a common *jokakwaro*. To say this means to stress common descent from an apical ancestor and, thus, to stress a sense of linkage with others.

18 In some respects, Rarieda was unusual, because several *jokakwaro*s had a number of members with secondary, or higher education and they held quite highly paid jobs. For the same reason, an unusually large number of migrants from Rarieda were clerks or supervisory workers, rather than artisans or unskilled workers. This made Rarieda interesting for a study of class formation, but it also made it important to cover other lineages. I, therefore, also drew my budget survey sample from members of Keyo and from JoKorwa, a large, generally poorer and less educated *jokakwaro* within JoKatombo which has settled adjacent to Rarieda. In order to construct a sufficiently large and varied sampling frame for the sample survey, I carried out a census of all households in Rarieda, Keyo, Korwa and Kadero. This provided a sampling frame of 373 households, from which I chose 100 households randomly. Four households in the sample were later found to be newly formed households not yet farming separately and so a further four households were chosen at random.

19 Whisson 1964: 33. Whisson also stresses the importance of cross-cutting ties, especially those traced through women. Affines often helped out in times of need.

20 It will become clear later that a fixed distinction between male- and female-headed households often cannot be sustained (cf. Chant 1997). This table simply gives a rough indication of the range of household structures.

21 The sample was not random. It was chosen to cover individuals in different types of household, in order to capture the widest possible variety of household types. The initial dimensions of variation were generation, migrant male head/non-migrant male head/female head, and better-off/poorer.

22 Derived from Table 3 in Francis and Hoddinott 1993.

23 *Riembo* (lit. 'to drive away') is a fostering relationship. An individual with livestock gives them to another person, on the understanding that they can eventually be claimed back. The recipient does not own the livestock, but has the right to keep the animal's offspring and use its products. This arrangement allows a household without enough grazing land to own livestock and it gives households without livestock a chance to build up their holdings. *Riembo* is also a tangible sign of friendship.

24 In a random sample of 104 households, 34 per cent sold beans in 1988, 33 per cent sold green vegetables, but only 7 per cent had sold maize.

25 Studies of urban migrants from Western Kenya include Parkin 1975, 1978; Grillo 1973; Southall 1975; Moock 1976; Ross and Weisner 1977; Cohen and Odhiambo 1989, 1992; White 1990.

26 Anker and Knowles (1981) using a more broadly based sample, found that only 6 per cent of income was remitted. Johnson and Whitelaw's figures (1974) are more relevant to this analysis, which is concerned with the impact of falling real wages on the remittance behaviour of middle and lower-income groups, see also Mukras *et al.* 1985.

27 Jeremiah's tone when describing these decisions made it clear that they were his.

6 Migration, accumulation and poverty

> What is lacking [in the historiography of Kenya] is the history of the 'raia' [ordinary people]. In the constellation of lives that constitute the Kenya of the colonial and postcolonial eras, there are not only the faceless and nameless 'raia' and the prominent and privileged … there are many who live seemingly uneventful lives as artisans, women working their rural lands, peasant farmers, transporters, labourers – whose lives marshal the interest of historians and other record keepers only as they educate their sons all the way to Harvard. … Even then, recognition most often comes not to the struggling 'raia' who produced those who achieved visibility and prominence, but rather to the achievers themselves.
>
> (Cohen and Odhiambo 1992: 43–4)

> My sons say, 'I'm coming, I'm coming', but they don't come home. I cry out to them, 'What will you do when I die?'
>
> (See the life history of Peter Olal, p.129–30)

Labour migrancy has shaped the lives of everyone in Luoland over the last century. Most men worked as migrants at some time in their lives; many women moved to town to join their husbands or to find work. People staying behind had to cope with the absence of their relatives for much of the year. Many depended on remittances for their livelihoods. Yet Luo people's experiences of migrancy have been very different. Some moved into Kenya's new professional middle class, living in towns, and their children part of a globalised youth culture. There was a window of opportunity for rapid mobility around the time of Independence, but most people missed out on it. Instead, they faced low wages or unemployment in the towns, or an equally precarious life in the countryside, living off a tiny farm and dwindling remittances. This growing class division reverberates through popular culture and the issues it raises are painful – can you have a real home in Nairobi, or only in Luoland? What link, if any, is there between young, middle-class Luos growing up in Nairobi and poor farmers from the same lineage in Nyanza?

In 1987, these issues were dramatised in a struggle over the burial place of S.M. Otieno, a prominent lawyer. His widow, a Kikuyu, wanted to bury him on the farm he had bought near Nairobi, while members of his Umira Kager clan insisted that he be buried at his birthplace in Siaya District. Otieno's body lay in the mortuary for six months and the case went to the High Court. The clan won the case and Otieno was buried in Siaya. The arguments heard in court, in the press and all around Kenya ranged over many issues and it would be over-simplifying to say that the case was really about class, gender, ethnicity or modernity.[1] Its class dimensions were clearly articulated in the disdainful atti-tudes Otieno's widow and sons displayed towards their rural relatives and they were also implicit in the anxiety Otieno's less successful kin showed to claim him as one of their own. One of the reasons why class is such a painful subject is that pre-colonial Luoland seems to have been a relatively egalitarian place. There were differences in wealth and power, but there was no hereditary ruling class and social boundaries were permeable (Whisson 1964; Hay 1972; Pala 1977; Butterman 1979 and oral evidence). This egalitarian legacy is still visible. There is none of the automatic and stylised deference towards the powerful one finds in more hierarchical societies.

Differentiation is a central part of the story of social change in Luoland, but it is a complicated story. There are several reasons for this. In the first place it is difficult to see the full picture from the rural end of a migrant labour economy. In the early part of the narrative that follows, stories from the generation born before 1930 tell how some people accumulated resources, achieved upward mobility and managed to transmit some economic advantages to their children. By the 1980s, the only significant accumulation was going on in the cities. The children of the successful people from the older generation were almost all living elsewhere. The position of people now living in Koguta was not just the most recent instalment in a linear story of differentiation. The change of gear in the narrative, between stories of accumulation and upward movement and stories of impoverishment, reflects a disjuncture in the social history of Western Kenya. Another gap in the narrative stems from the fact that it was men who got education and better-paid work in the colonial period. For this reason women's life histories, and gender relations, remain in the background in the early parts of the account that deal with processes of upward mobility.

There are other challenges to our understanding. One concerns units of analysis. If differentiation has been so important, between what groups has the most important differentiation happened? Between lineages? Between compounds? Between houses? Between individual people? Between genera-tions? By the late 1980s, most people's livelihoods relied on strategies organised around households, or they relied on their own efforts. Differentiation between households was the most significant kind of differentiation. However, reading backwards from these strategies onto the strategies people followed in the past would be misleading. There is quite a lot of evidence to suggest that relation-ships beyond the household were more important to the livelihoods of earlier generations than they later became. They gave some people access to the labour of others and to land and livestock. People's increasing reliance on markets for their means of livelihood weakened the relationships of authority or reciprocity on which this access rested. Yet it would also be wrong to conclude that now only households mattered. In the late 1980s, some people relied on kin and friends to keep them away from destitution; many more could look outside the household when they needed help. What was much less common was day-to-day co-operation in production. Relationships beyond the household also affected long-term patterns of differentiation. Middle-class JoKoguta living in Nairobi were mostly JoKomoro. Their fathers and uncles had embraced Christianity and mission schooling in the 1920s, giving the children of their lineage a headstart on their more distant relatives. But many JoKomoro had not got caught up in this process of upward mobility. The transmission of economic advantage within lineages had happened, but it was by no means automatic. Making sense of social change involves sometimes focusing on individuals, on the choices they made or the strategies they followed. At other times, it involves looking at households and the ways in which people's decisions are affected by their relationships with other members of their households.

Another challenge is to decide how much differentiation is a cumulative process, rather than the working out of households' developmental cycles. The concept of a household developmental cycle was a central element in Chayanov's theory of demographic differentiation within a 'peasant economy' (Chayanov 1966). In contrast to Lenin, who argued that Russian agriculture was essentially capitalist and that differentiation was polarising agrarian society between proletarians and a bourgeoisie, Chayanov argued that Russian farm households were operating within a non-capitalist 'peasant economy'. Chayanov's model of the peasant farm household assumes utility maximisation (the household makes a trade-off between drudgery and the income needed to feed the household, rather than maximising profit) and the absence of wage labour. Because all labour is family labour, farm production is determined by the ratio of consumers to producers in the household. This ratio changes over the household developmental cycle (as children are born, become old enough to do farm labour and then marry and leave), and farm income varies accordingly. Chayanov and his associates argued that cyclical mobility explained most of the differentiation which Lenin perceived to be evidence of polarisation (Chayanov 1966).

At first sight, Chayanov's theory does not seem to have much to offer the analyst of an economy where the vast majority of men have for decades spent most of their working lives in wage labour outside their households and Kitching dismisses it.

> To show that, in a particular set of circumstances, the autonomy of the household as a production and consumption unit does not hold, and that patterns of differentiation can indeed be cumulative because household income is not primarily determined by the household's agricultural labour resources, is to successfully demonstrate the irrelevance of the Chayanovian model to that particular set of circumstances.
>
> (Kitching 1980: 63)

However, it is possible to separate the concept of a household developmental cycle from the assumptions of Chayanov's analytic model of the family farm and, indeed, incorporate links between farm households and the wider economy into an analysis of developmental cycles. In a migrant labour economy, this involves two challenges. The first is to understand how households' access to resources, including land and wage income, change over the developmental cycle. The second challenge is to establish whether differentiation in access to land and wage income implies that different households in fact face quite different developmental cycles (cf. Spiegel 1980 and Murray 1981 on Lesotho). This is difficult to determine and can probably be fully understood only through a study of long-term changes (Peters 1983). Even then, it is extremely difficult to reconstruct household structures in earlier periods (Vaughan 1983). Longitudinal studies, where feasible, would provide one solution to the problem, but are normally impractical on the time scale needed to see long-term trends. Life histories are a rough-and-ready substitute.

'The men who were keen on education didn't want hunger': upward mobility and the transmission of economic advantage

In Nyabondo, the African Inland Mission set up the first schools in the early 1920s, but their enrolment was small. Many parents were suspicious of the missionaries and they wanted their children at home to look after livestock. The first children from Koguta to go to school came from the families of the early converts to Christianity. Adopting Christianity brought along with it a commitment to education. The early Christians became aware of how important it was in the new colonial economy ('The men who were keen on education didn't want hunger', interview, James Onditi, 12 December 1988). Many of the older men from Koguta who received a few years of primary schooling belonged to the same *jokakwaro* as Philip Ojwang. Ojwang was an early convert to Christianity who worked for the colonial administration, eventually becoming Vice President of the local African Court, and he encouraged men in his *jokakwaro* and their *libembini*, who lived in Rarieda, Jokomoro to send their children to school. The fact that so few Africans had even a few years of primary education put a high premium on education once young men started looking for work.[2] A young man who had a few years of schooling could find work as a clerk or in a skilled job (Lonsdale 1964; Whisson 1964). In the 1930s, young men with primary schooling worked as teachers, clerks, field assistants or in the African courts. In the 1930s, 1940s and 1950s, most of the better-paid migrants were Jokomoro, or were from closely related lineages: a head start that has shaped patterns of differentiation between lineages ever since. People in the same lineage not only encouraged one another to educate their children. Some Jokomoro who already had better-paid work helped their younger *joanyuola*. A *wuod Komoro* or a *nyar Komoro* (son or daughter or Komoro) in a steady job would be expected to, and often did provide money for school fees, accommodation in town and help in job seeking. They would do this, not only for their younger siblings and half-siblings, but also for other young relatives, including their affines. The results were still clear in the late 1980s. Most JoKoguta who were urban professionals, or indeed who were in clerical jobs, were JoKomoro. This pattern, where one lineage in a sub-location did much better than everyone else, is common in Luoland. If you walk through a sub-location you will find one area where there are more permanent houses (cinder block or brick) or semi-permanent houses (mud and cement) than elsewhere. This will be the lineage who first converted to Christianity and sent their children to school.

Peter Olal was born around 1910 at Sango. His father died when he was a child and his mother brought him to Koguta when he was somewhere between eight and ten years old. Philip Ojwang (a senior man from the same *jokakwaro*) gave her a field and looked after her. Peter went to Bwana Innes' mission school for two years. 'In those days, not many children went to school. The children who went were those who had connections with the Church. Not many children from Rarieda went – that came later, with their children. Ojwang and Mboya were very involved with the Mission. They were good leaders and they told the men here to send their children to school.'

After he left school, Peter went to Muhoroni to work as a domestic servant, staying for two or three years. Then he came back and worked on his mother's field and herded her cattle. She had a lot of cattle that came as bridewealth for her daughters. When Peter was in his mid-twenties he married two young women in the same year. He then went to the Veterinary School in Maseno for two years before starting work as a veterinary scout in 1933, a job he kept until he retired in 1973. He lived and worked at various places in Kisumu District and the Rift Valley – Ahero, Koguta, Songhor, Nakuru and Eldoret.

His wives would take it in turns to live with him, swapping over every three months. They grew maize, sorghum and millet. They did not sell grain, but they did sell beans and sweet potatoes. Peter bought a plough in 1942. He came home and did the ploughing in December and he used to visit for three days at a time in the weeding season. He employed a herdsman who would work the plough team when Peter hired it out. His wives employed a little labour, 'but it was hard to find people to work for them, because there was plenty of food in those days'.

Peter had a lot of land (thirty hectares, much of which he had acquired during the land consolidation in the 1960s), but most of it was lying idle in 1988. He had taken up coffee growing in 1975, but it was not remunerative enough to make a big investment worthwhile. He was now too old to do any farm work and his main source of income was his pension.

Peter's main investment was his children's education. His only other investment away from the farm was to build a shop, which he rented out. He was not highly paid. He often could not afford to send money home to his wives. But he built up a herd of cattle (his veterinary knowledge helped) and he gradually sold them off to pay his children's school fees. Two of his sons were university graduates. One had a high-ranking job in local government in Nairobi and the other was a manager in a multinational company. Two daughters were at university in 1988. Two other sons were doing clerical work and another daughter was a nurse in the UK. The younger children had not been so successful. Two sons went to secondary school and were clerks, another son had come back to live at home.

Peter felt that that his children did not visit often enough and that their commitment to their home was weak. 'They keep promising to build good houses here, but they don't (they had built mud-and-cement houses). My sons say "I'm coming, I'm coming", but they don't come home. I cry out to them,"What will you do when I die?" Many wealthy Luos just stay in town, because they have built houses there.' In fact, Peter's sons had made some investments. The manager in the multinational had started a posho mill and a paraffin pump. But his other investments were elsewhere. He and the local government officer had both bought land in the Sugar Belt and built rental housing in Nairobi and Mombasa. Peter's children were still giving him support, however. They sent him remittances totalling KShs.1000 in the year 1988–89.

John Olal was Peter's cousin. He was born around 1915, also at Sango. An orphan, he was fostered by Peter Olal's mother and was eventually given land by Philip Ojwang. He studied the 3Rs at Nyamaroka and then went on to Nyakach mission, in 1931. After three years, he took some examinations and then came back home while Peter went off to train as a veterinary scout. He then took the same course himself from 1937 to 1939. He worked as a scout for a year, earning about 10 shillings a month, but in 1940 he got a job as a clerk on a coffee farm at Sotik, in Kericho District. His employer was called up, so in 1941 he moved to Brook Bond at Kericho, starting on 30 shillings a month. That was good money. He married his first wife in 1937 and his second in 1948. His first wife lived in Koguta and his second wife lived with him in Kericho. Slowly, he was promoted and eventually became a senior clerk when Africans replaced Asians after Independence. He paid into a company Provident Fund and he got a lump sum when he retired in 1971.

After John retired, he set up a hardware business with a partner in

Sondu. He did this for five years and then he stopped, because he was getting too old. He used the proceeds from this business to pay school fees and his older children also helped out with the fees of the younger ones. He had sixteen children altogether. They have had very mixed experiences. One daughter went to university. She took a postgraduate degree and became a statistician. John's eldest son went up to Form Four in secondary school and became a manager in a bank. Several daughters were doing clerical or secretarial work. Others went to secondary school but were not working. The younger ones had not done so well: some were looking for work. The two youngest were still at school. (This contrast between the older children, born in the 1940s and 1950s, and the younger ones reflects the very different conditions under which the latter entered the labour market. Nobody getting a Form Four qualification in the 1980s could hope to become a bank manager.)

People in other lineages, whose members were more suspicious of mission churches and schools, converted later to Christianity. They caught on to what was happening and, from around the 1950s, started trying to give their children enough education to improve their chances in the urban labour market. But, doing small-scale farming or working in low-paid jobs, few of them had the resources to educate their children beyond primary level. Primary education was no longer a guarantor of better-paid work and their children mostly working in low-wage urban jobs or trading in the informal sector. Though most JoKoguta with white-collar jobs were JoKomoro, the majority of JoKomoro migrants were not doing white-collar work. Many JoKomoro living in Koguta were poor. Some were *jodala* with the urban professionals. Differentiation within lineages, and *mier*, was as significant as differentiation between lineages, and often more so. The life histories also show how much differentiation there has often been between older and younger siblings.

Men who worked for many years did make some investments in their farms. Peter and John Olal both built up herds of cattle and Peter bought a plough. But their priority was their children's education. Peter sold cattle to pay school fees and John did the same with the proceeds of his hardware business. There were many reasons why men who had better-paid jobs were reluctant to put many resources into farming. Unlike Central Province, close to Nairobi and the European farms, Luoland was far from major markets and it was plagued by the poor state of communications. None of the cash crops that Africans were allowed to grow brought a return that was high enough. Yet the difficulties cannot have been overwhelming. A group of slightly older men had with less or no schooling made a success of trading and small-scale cash cropping and did not have to look for work.

James Onditi was born around 1906 at Sango. His father was a very keen farmer, but he died while James was still a baby. His mother was inherited[3] and she and his father's brothers moved to Koguta. 'There was a lot of death at Sango and people were marking out the land up here.' James went to Bible school in 1924 but he learned 'only a little'. From 1926–28 he worked on a sisal plantation. 'I went because I wanted to buy clothes. I stopped working there because there was not much food at home. I wanted to farm. I did not go out again because I was a good farmer and because I was trading in cattle.' James married in 1930 and bought a plough in 1939. His mother died in 1933 and he was given four fields. He also sharecropped three more fields. In the 1930s, he would get about twenty sacks of maize from this land and he sold three or four of them.

He took cattle from Oyugis (South Nyanza) and sold them in the market at Ahero. He traded cattle for about five years. When he had sold them all, he turned to trading in grain, taking maize and millet by donkey from Kisii and Narok to Ahero and Kibos. He also had a hand-gristing mill and people would bring their maize to grind. 'I was like an educated person – I got more money than them.'

James joined the first Africa Inland Mission and later the (Anglican) Church Missionary Society. He was one of the men who built a school and church at Guu in 1931. 'Our working outside taught us to build the church and the school. I would walk around and encourage people to school. JoKadero liked alcohol and *bhang* (marijuana), but we liked God. They were mean and didn't want to pay school fees.' James sold a lot of cattle to educate his children. Two went to university and have become prominent Kenyans, members of the professional and intellectual elite.

Men like James took advantage of the limited opportunities that did exist, selling their crops and trading.[4] In theory, migrants could have done this as well. Their wives were living at home and could have expanded crop production, especially if they had been sent money to pay farm labourers.

Even the migrants who could have afforded to did not follow this strategy. Most of them were too far away to be able to visit often, so they could not have supervised the farm operation themselves. Many were reluctant to delegate to their wives enough authority to make the decisions involved in expanding production on the farm, as we shall see in Chapter 8. Migrants in better-paid work tended to send home only enough money to supplement the food supply and buy basic household goods. When larger amounts of money were needed, for clothing, school fees or building work, the husband either paid for these himself or earmarked the money beforehand.[5] These men made a strategic judgement that farming was never going to bring much of a profit and that they could not combine it with labour migration. Some of them did put time and money into farming after their retirement, but only as a fallback after other enterprises like transport and trading had failed. Men who had enough grazing land preferred to invest in cattle. Most felt it was more important to invest in assets like rental housing, with fixed returns. More than anything else, they put resources into their children's education. They sold their cattle and diverted money from other uses to pay school fees.

People hoped that children would eventually support them. Better-paid migrants, local traders, craftsmen and farmers alike invested mainly in their children's education, in the self-fulfilling prophecy that this would be the only route to upward mobility for the next generation. These children were a small and privileged minority. An indication of just how small this group was comes from a 1937 Education Department Report, which gives the following enrolment figures for Africans. Elementary schools offered 4–5 years of education; primary schools offered 8–9 years.

Table 6.1 Africans enrolled at school in 1937

African schools	Elementary	Primary	Secondary
Government	3,175	1,218	–
Non-government*	90,474	8,139	273

Source: Adapted from Kenya 1937

Note: *plus 7,223 pupils at independent schools

Few men bought land locally on any scale. Some did acquire quite large amounts of land through expropriation, either through litigation or through their work as land elders when land was consolidated in the 1960s, but they did not develop it. Others bought small plots locally, usually because they wanted to pass on land to their sons, rather than to invest in farming. Men who were interested in commercial farming and who had the resources generally bought land in the Settlement Schemes in the Nyanza Sugar Belt. Larger tracts could be bought there than in Koguta; the land could be bought without fear of its later being reclaimed by the vendor and it was suitable for sugar cane production. Living there also gave some insulation from poorer relatives and their requests for help.

As school enrolment rose in the 1930s and 1940s, primary education became more common. By 1954, 312,597 pupils were enrolled in African primary schools in Kenya and there were 45,295 pupils in African intermediate schools (Kenya, Colony and Protectorate of 1954). This change increased competition in the labour market. As more people achieved basic literacy, a secondary-school education became essential for finding clerical work. The small minority of people from Koguta with secondary and university education, sons and daughters and wards of people like Peter Olal and James Onditi, were well placed to take advantage of the opening up of professional employment to Africans in the 1960s and 1970s. But their ties to their rural kin were weak.

Their attention was centred on the urban economy. When they did become involved in farming, it was on land they bought in the Nyanza Sugar Belt. Nor did many of them send much money back to Koguta. Some children had built their parents permanent houses, but they were a minority. These parents' expectations of regular support had mostly gone unfulfilled and many had a strong sense of grievance.

After Independence, secondary-school numbers increased very quickly. Collier and Lal estimate that between 1945 and Independence 'the total output of African secondary schools must have been slightly less than 8,000. This contrasts with primary school output approaching 27,000 per annum in the immediate post-war period and around 100,000 per annum just before Independence' (Collier and Lal 1986: 59). The stock of Africans with Form Four education rose fourfold between 1963 and 1969 and it continued to rise rapidly, with African Form Four enrolment doubling in 1967–68 and increasing 3.2 times between 1968 and 1978. Collier and Lal estimate that the chances of an African boy gaining a Form Four secondary education increased sixfold between 1964 and 1977, from something less than one-in-twenty, to something better than one-in-four. This massive rise in numbers quickly produced an excess supply of Form Four leavers in the period 1967–72 and since then a Form Four certificate has been no guarantee of formal sector employment, while some form of higher education has become a prerequisite for most professional jobs other than teaching. This massive increase in secondary-school numbers was largely due to the creation of self-help *harambee* secondary schools after Independence. In 1977, there were 268 primary schools and 23 secondary schools in Kisumu District (Kenya 1977). In the 1980s there were two secondary schools in Koguta alone, with a secondary school added at Guu and a new secondary school at Naki. As well as the many *harambee* schools in the area, there were also two government secondary schools on the Plateau, Nyakach Girls and Nyakach Boys. The returns to investment in secondary education (and university education) became lower and increasingly uncertain. This meant that even migrants with stronger rural links could not afford to send much money back.

The case of Odiyo shows why men who followed a strategy of accumulation through farming alone were not able to provide their children with a stepping stone into the professional middle class.

Peter Odiyo was born in Seme in 1914, the eldest son. He didn't go to school. 'When I was old enough, I decided that I did not want to be a *jadak*, so I came to join Kadero people here in 1936. All the land had been taken, but my clan-mates pushed out the original occupants of this land. A clan-mate claimed the land here, so I had to be strong enough to push him out. I needed money to pursue the case and so I began trading in cattle. I stopped in 1938, once I had enough money and influence.

Then I bought a plough and settled into farming, which I have done ever since.' In 1988, he claimed to have over sixty acres of land, divided between his three wives and his own *mondo* (husband's field). His wives did all the agricultural work, supplemented by the labour he employed on the *mondo* and they employed on their fields.

Odiyo explained that he sold a lot of his output, but had no other investments. In 1987, he got 40–50 bags of maize. He also grew sorghum, beans, oilseeds, sweet potatoes and bananas. He usually got about 20 bags of maize from his mondo and sold them all. He also took five bags from each wife and sold them. He also grew oilseeds for sale – ten bags in 1988 – and he grew eucalyptus and thatching grass on the mondo, which he also sold.

Odiyo had twelve sons. The first and second were teachers, others were a mechanic, a carpenter, a clerk and an agricultural assistant. Five other sons had a Form Four certificate, but were unemployed. (Odiyo's wives also mentioned their daughters. One had gone up to Form Four and was working in the Ministry of the Environment). Odiyo's sons sent him money to pay for the younger children's education and he also raised the money by selling cattle, maize and oilseeds. He thought he was paying out about KShs.15,000 a year.

This story shows how much more competitive the labour market had become. Odiyo's children went to secondary school, but only the oldest ones found professional jobs, as teachers. Even they had not done particularly well. Teaching was a badly paid and low-status profession by the 1980s. The younger children had simply joined the ranks of unemployed Form Four leavers. That is not to say that all the hard work and conflict had been for nothing. It was better to have two sons who were teachers than see all the children unemployed, but the story shows how hard it was for farmers to fund their children's education for long enough to allow them to find a secure job.

In households which did manage to give their sons and daughters enough education for them to get professional jobs, the children made very few investments back in Koguta and their links with the rural area were weak. Isaac Otieno was unusual, but even he had to put most of his time and energy into his job in Nairobi.

Isaac Otieno, born in 1948 was a consultant, managing East African trials for an overseas agro-chemical firm. Isaac's father was a 'semi-literate' trader. Through his trading, he made friends with Asian and Nubian traders. 'They advised him to forget about cows and concentrate on education.' So he sold his cows to pay school fees. Isaac had been educated up to Master's level and had studied in Canada. Isaac had not been home for longer than four weeks since 1960. One of Isaac's brothers was General Manager of the Kenyan branch of a multinational company, one was involved in fisheries research, one was a teacher and the other was living at home. One sister was a farmer and the other was a teacher.

Isaac was divorced and his three children lived with their mother. He never saw them, but he hoped that his sons would one day come and build a *simba* in the *dala*.[6]

(Isaac was interviewed when he visited at Christmas. Not many other migrants had come to visit.) 'These days, people cannot afford to come home on leave.' One of his friends from Koguta put it this way: 'He is a teacher in Mombasa. People say that you get lost in Mombasa, but just to get himself and his wife, let alone his children, to Kisumu from Mombasa, and back, costs KShs.800. People now come home only for emergencies.'

Isaac was Chair of his lineage's social welfare organisation in Nairobi. He saw three types of people from Koguta in Nairobi: 'At the bottom is the mass. They live in cheap housing; the wife is here and is quite happy; and the kids go to school at Guu. Coming home for this type is no problem – he can just jump on a train. Then there is the semi-elite – people like me. They are up to their necks, because their families are with them in Nairobi. There is a lot of expense involved in coming – travel, food, paying for a watchman for the Nairobi house. Then there is the elite – they can take their families on holiday to Mombasa and come home quite easily.'

The older generation's expectations were unfulfilled because the shift in spheres of accumulation from farming and rural activities to the urban labour market that took place during their lifetimes had become almost complete. Isaac was extremely unusual in trying to run a farm. It would be almost impossible to make a profit by putting money into investments in the area. Where someone might think of making an investment (such as in the transport business), it was difficult to supervise at a distance. The wide gap in incomes between urban professionals and their rural relatives was another problem. Most people would prefer to avoid constant demands for help.[7] So they made their investments outside Koguta. They built houses to rent out. Like the older generation, they bought farmland in the Nyanza Sugar Belt. Neither investment strengthened their ties to Koguta in any way.

The links between Luo urban professionals and their rural relatives were becoming more and more only symbolic. People reasserted their membership of the lineage by coming back for funerals. Old people pressured their sons to build houses, to show commitment to the *dala*. But these houses were locked and empty for all but a few days each year (cf. Goldenberg 1982). Many of their parents had fallen into poverty in old age. The households in the most fortunate position were the long-term migrants who had pensionable jobs, but their incomes were still low. Far from being able to sustain accumulation, they often could maintain a reasonable standard of living only by selling their assets. Cattle were sold off. Land went untilled. Old age and ill health reduced their input into farming and they did not have enough money to employ much labour or make labour-saving investments.

These men felt a real pride in their achievements. By ingenuity, hard work and good fortune they had managed to get their children into Kenya's small class of professionals, *Josomo*. But they themselves were facing an uncertain future. They could not avoid the pressures of the household developmental cycle, because they did not have the resources to overcome the problem of declining household labour power. Some tried to get around the labour shortage by marrying a young wife, but this raised a further difficulty of feeding another household and, eventually, educating the children. Young women in this position might start trading or small-scale cash cropping, but their children had a only a small chance of achieving an economic position comparable to that of their older half-siblings, unless these financed their school fees. This did happen. Urban professionals provided support for younger siblings, half-siblings and more distant kin. But the cost of education was rising and it was becoming difficult to keep up this kind of support. For this reason, younger members of the same *dala* and *jokakwaro* as the urban elite had completely different life chances. They might get a few years of secondary education, but this might not be enough to secure any kind of formal-sector job, let alone a foothold in the middle class.

Migration and poverty

In lineages other than JoKomoro people were more doubtful of the benefits of education when the missions set up schools in the 1920s.

> 'People said that those who went to school would get scabies ...'
>
> (Miriam Okongo, born *c*.1906–12)

Sending children to school meant taking them away from herding and other farm and household work for no immediate benefit. Some people also felt suspicious of the missionaries, while young people themselves fell foul of the latter's strict morality.

> 'I had two years at Bwana Innes' school, until I was sent away for talking to girls.'
>
> (Wilson Okongo, born 1904)

Almost all of the young men in these other lineages who were born before 1930 spent a period at work outside Koguta, but their patterns of employment were markedly different from those of their contemporaries who had more schooling. Most were target agricultural labourers, but some men who began in this way then worked outside Koguta for many years, gradually undertaking more permanent and better-paid employment.

> Ondiek was born at Kano in 1910. His family were *jodak*. They came here because they did not have enough grazing land by the lakeside. He went to primary school for about seven years. In 1928 he went to Kericho to work as a tea packer. He would work for a month or two and then come home. 'We still liked home.' He earned 3 shillings a month. 'We would buy cloth and when it was torn we would go back.' In 1934, Ondiek got a job at a golf club, working in the store. He eventually became the overseer, earning 200 shillings a month. He also did a little work in the office. He returned home for good in 1970.

But others, after a few years in which they earned bridewealth, returned for good. Many men of this generation and level of education were quite strongly oriented towards living in Koguta:

> Michael Obiero was born in 1906 at Thudibuoro and his family came here when he was about five years old. They came because they wanted land. He went to Bwana Innes' school. During the First World War, he was sent to do forced labour, building houses. He did this for six months, earning 12 shillings a month. 'That was good money. You could buy a cow for less.' He then spent some time as a squatter at Muhoroni with his father and some uncles.[8] He stayed for six years. Eventually his relatives persuaded him to return to Koguta. In 1934, he went back to Muhoroni and became a foreman in a sugar factory. He did this for a year. Then he quarrelled with his employers and went to Kericho, working as a foreman in charge of 12–13 tea pickers. He did this for two years, and then he came home for good.

If a man's prospects on the labour market were not good, and if he could hope to earn money locally, he might decide to stay in Koguta. Some of the men who did this had acquired skills like masonry when they had been migrants, or they might find one of the few jobs available. Others had just enough land and cattle to satisfy their food and cash needs from farming. People like this did not manage to find profitable niches in the local economy, however. Their incomes were lower and they could not invest much. Other men did not have the option of staying in Koguta. They had too little land or no skills that could bring in an income. They had to keep on working on European farms or in the towns. Some were away for as much as forty years. They struggled to send back remittances that were used to buy clothing and household goods and to pay school fees. Because these households had only small plots of land for farming (even at that time, some had as little as three acres), fallow periods became shorter and often disappeared altogether. The continuously cropped land quickly lost fertility. Falling yields meant that output in many households eventually dropped below subsistence level.

Some of the food deficit was made up by a withdrawal of labour from agriculture. Women spent less time in the fields and, instead, traded in fish or other food, or sold the palm frond baskets which became a regular feature of the local economy in Nyakach. These shifts into petty trading and petty commodity production set off a downward spiral of falling output and falling labour input into farming. This called for an increasing proportion of household income to be found from off-farm sources. Remittances were needed to pay for food. Households became locked into a pattern where men migrated and women stayed behind in a holding operation.[9]

Many people in this generation judged that the best way to avoid impoverishment in old age was to make sure that their children had secure employment. It was a vain hope. As we have seen, they would have needed secondary education, which had always been much more expensive than primary education, to benefit from the expansion of professional jobs when they came on to the labour market in the 1950s and 1960s, still less the 1970s and 1980s.

Rising real wages in the 1950s and 1960s benefited skilled workers (Collier and Lal 1986), but few of them invested in agriculture. Instead, they found better housing and brought their wives to live with them in town. If the husband was a polygynist, his wives might take turns to stay with him.

Ruth Ogendo was born in Kangan (Nyabondo) in 1948. She was eighteen when she got married. When she married Ogendo, he was working as a guard for Kenya Railways, but he was now a ticket examiner. Until he married his third wife, she lived with him outside. She would come home for one month a year and then go back. They lived in Nakuru for seven years, then in Gilgil for four years. Then they went back to Nakuru for two years and to Nairobi for six years. When Ogendo married another woman, Margaret, they would swap over about every three months. Ogendo married another wife in 1984. Now this woman lived with him all the time.

Men with lower incomes also followed this pattern, but, for many of them, the wife lived in town for a shorter time, or her stay was punctuated by long spells in Koguta. Produce from the farm in Koguta might be a useful supplement to the household's income in town.

> Grace Akinyi was born in Kadiang'a (the sub-location next to Koguta) in 1935. She went to primary school up to Standard Four and she married in 1952. When they got married, her husband was working as a carpenter in Thika (near Nairobi). In 1962, he came back to Koguta for a year and a half. Then he went and worked as a carpenter in Kericho. He finally came back in 1975. 'When he worked outside I lived with him. I would come here to plant and weed. Then I would stop and go back. I did this because I wanted to avoid problems in getting food. When I was outside, I would do a bit of trading in vegetables. People would come to the house to buy. When I was at home, I would trade in beans and vegetables every market day.'

Once there were several children to feed, it was cheaper for the wife to move into the *dala* and start farming.

Coming to live in her husband's *dala* has always been a major change in a woman's life. Before women began to spend long periods living in town, a young woman moved into the *dala* as a newly married wife. Increasingly, women came to live in the *dala* permanently after several years of marriage and arrived with several children. This change had profound implications for their relationships with the other households in the *dala*.

Winjruok and nyiego: relationships in the dala

The coming of large-scale labour migrancy made the *dala* far less of a unit. The authority of the *wuon dala* was weakened by the fact that his sons no longer depended on him, or his fellow elders, for bridewealth cattle. They could earn the money to buy their own. In many homes, the flow of resources was reversed. Old men and women needed money from their migrant children. Senior women in a *dala* began to have much less control over the labour of junior women. Until the 1950s and 1960s, when many young women began to live with their husbands in town, a newly married woman was expected to spend the first few years of marriage working in the fields of her mother-in-law, or one of her husband's senior wives, and cooking with her. When a woman with young children moved to Koguta after several years in town, however, she would usually expect to be given her own field straight away. She would already have her own load of domestic responsibilities and little time to spare for her mother-in-law. The loss of men's labour power also meant that younger women had less time available to work for their seniors.

These trends were not confined to farm labour. Households in a *dala* would help one another out in a domestic crisis, or when there were visitors. Whether or not they did more than this depended on whether there was *winjruok* ('understanding', 'empathy') in a *dala*. But even then co-operation was usually confined to occasionally sharing childcare and shopping. The labour burden for each household was too large. Spare time from domestic work was taken up with trying to earn money. Households within a *dala* were more likely to share food than to share labour (see the discussion of transfers between households on pp.152–6).

The greatest threat to *winjruok* came from *nyiego* ('jealousy'). *Nyiego* (note the similarity to *nyieka*, 'co-wife') is the word people reached for when describing an unhappy *dala* (cf. Parkin 1978). There were many reasons why *winjruok* might be more vulnerable to *nyiego* than previously. Because land was becoming so scarce, there might be competition for land within the *dala*. Overt conflict over land was still rather rare, but most people could think of cases. There might also be tension if some sons had found better jobs than others.

As well as causing tension, the tendency for each household within the *dala* to become a separate unit of production had detrimental effects on the welfare of old people. Very few of them were sent enough money to replace the lost labour of younger women with hired labourers.

Migration and differentiation: conclusions

The story of socio-economic differentiation in Koguta is disjointed because the processes it describes cannot easily be fitted into a continuous narrative. Nor can they all be captured in a rural-based study. The early part of the narrative describes accumulation, upward mobility and the transmission of economic advantage between the generation born before 1930 and their children. Because of a series of historical disjunctures, these themes later disappear. The position of people living in Koguta in the late 1980s was not simply the outcome of linear processes of differentiation. The formation of a class of urban professionals was rooted in the incorporation of Koguta into the colonial economy. Parents paid school fees with money earned from migrant labour and from involvement in the new food and product markets that opened up. Differentiation in the rural economy was slowed down in favour of a rural-urban polarisation that was still growing. Large-scale accumulation and differentiation centred on access to earnings and profits in the urban economy. They were the province of people based in the urban areas.

Notes

1 The court proceedings, and the national debate they triggered, have been widely discussed (Ojwang and Mugambi 1989; Cohen and Odhiambo 1989, 1992). Cohen and Odhiambo argue convincingly that the case cannot be explained in terms of 'really being about' one issue or another.

2 Kitching shows that there were large and persistent wage differentials in the African wage labour force from the earliest years of the colonial period (Kitching 1980: 254–68).

3 Widow inheritance (leviratic marriage) was common in East African societies where a woman married into a patrilineage. A widow of child-bearing age would be expected to marry a senior man from her husband's lineage. Any children from the marriage would be raised in the name of the dead man.

4　As will become clear in Chapter 8, these strategies caused conflict over the appropriation of women's labour power and crops.

5　This was still the pattern in better-off households.

6　*Simba* – hut for young, unmarried men.

7　A small number of women born in Koguta found professional jobs and sent remittances back to their parents. People often said that daughters now helped more than sons. The data in the budget survey suggested that daughters might help more often, but that larger sums tended to come from sons. This reflected their greater access to money.

8　Whole families migrated to live as 'squatters' on European land around Muhoroni. They were given access to land in return for (at first minimal) labour on the farm. Squatters from Koguta moved in search of good grazing land. Many of the people who moved were quite wealthy, with large herds, rather like some of the early Kikuyu squatters in Central Province and the Rift Valley. Unlike other Kikuyu squatters, they do not seem to have been people with little land, or insecure tenancy, looking for arable land. Population pressure in Luoland was much less than in Kikuyuland. Nor were they looking for arable land in order to start farming commercially. Many of them returned to Koguta in the late 1920s and early 1930s, when European farmers increased their demands on squatters to supply labour.

9　Kitching suggests that, by the early 1950s, poorer households, particularly the households of long-term migrants, were likely to have sold land in order to meet short-term cash needs or because of a lack of labour to work the land effectively. Neither the evidence from the life histories nor the findings of a sample survey confirm this hypothesis for Koguta. 2.8 per cent of respondents stated that they had sold land. Informants unanimously agreed that it had always been difficult to buy land locally and that sale of land was something people resorted to only in the face of extreme pressure to raise cash. It was seen as an economically irrational act, as well as a great symbolic loss, because it severely reduced the vendor's income in later life, as well as that of future generations. Households had reduced the time they spent working on their farms, but household economies proved surprisingly resilient in the face of falls in output. Nor did better-off households buy much land. When they did buy land it was outside Koguta, in the Nyanza sugar-belt.

7   Poverty and livelihoods

There were two groups of people in Koguta. *Jonyalo*, 'people who are managing', were a minority. They were women whose husbands had a steady job and old men living on their pensions. The rest had to juggle different ways of finding money. Some people were better placed to do this. They might be women who could find some kind of trade that allowed them to tap into the flows of remittances coming into other households, or men who harnessed their wives' labour to grow crops for sale. Other people sometimes received a few shillings from their husbands or their children and the rest of the time oscillated between making baskets, working on other people's farms and selling a few pieces of fruit and vegetables.

A budget survey of twenty-four people living in eighteen households gives an overview of poverty and livelihoods. Tables 5.4 and 5.5 (p.117) give data on the components of income and expenditure. Table 5.4 shows how low incomes were in most households. Allowing for inflation, these low incomes are in line with the data from Kisumu District in the 1981/2 Rural Household Budget Survey. Differentiation in Koguta was now not about accumulation, but only about differing degrees of impoverishment. The important difference was between people who could maintain an adequate standard of living and the majority who were falling more and more deeply into poverty. People who received remittances regularly were just about managing on their incomes, as were people who could compensate for low remittances by earning money from off-farm activities. Remittances meant the difference between a regular, if low income, and a constant struggle to find money. Remittances were rarely large enough to allow people to use them for anything other than food, clothing and school fees.

Most households who received remittances depended on just one source – the state, husbands, children and other close relatives, mainly siblings. Of the eight women whose husbands were migrants, seven received remittances during the year of the survey. These remittances sometimes made up a very high proportion of their total cash incomes (over 80 per cent in three cases). Households receiving significant remittances from their children tended to be at a later stage in the developmental cycle: the husband might have died, or he had retired. Again, households whose children sent them remittances did not get much remittance income from elsewhere. However, two-thirds of the adult children (60 per cent of sons and 77 per cent of daughters) did not remit during the year. Children also tended to send money less regularly than husbands did.

Older people did not agree about what was happening to relationships between parents and children. 'They go to live in town and start to despise us', some said. They felt uneasy about this. Parents had to rely on their children for money, without having the authority to enforce their claims. Others disagreed. They could see how hard it was for their children to support their own households out of their low earnings, let alone their parents, and they did not feel that relationships had become strained.

The distribution of incomes in Table 5.4 is not simply the outcome of the processes of differentiation described up to now. Some of the households towards the top of the distribution, such as the Olals and Helen Ombura, were members of *mier* in the JoKomoro lineage whose older male members were early Christians who went on to get better-paid work. Others, like Millicent Mboya and Margaret Ogendo, while also belonging to these, or closely related *mier*, were in a much more difficult position. They had benefited from their close kinship with people who had been upwardly mobile. Millicent, now a widow, was the youngest wife of James Mboya. Mboya had been quite prosperous. He was a teacher, then a clerk and, finally, market master at Sondu. Several of his sons were university graduates. But Millicent had always had a struggle to find money. Mboya's resources could not stretch to giving her the financial support he had given his other wives and, although her children had some financial help from their half-siblings, they were much less well educated. Margaret Ogendo's husband, although doing a supervisory job with Kenya Railways, was one of the migrants being squeezed by falling urban real wages. He could not afford to send enough remittances to both his wives (Margaret and Ruth Ogendo) for them to avoid having to find money elsewhere. As the less-favoured wife, Margaret lost out. In contrast Elizabeth and George Owiny had no background of early conversion to Christianity and education, but they found a relatively profitable niche in food trading and cash cropping. Their position at the top of the income distribution shows the need to recognise that the processes of differentiation in operation were quite different from earlier.

The distribution of income was skewed. Most individuals, and households, had low incomes and there was a gap between them and the households at the top of the distribution. There were also quite large differences in the composition of different households' incomes. Elizabeth and George Owiny received 4 per cent of their income from remittances, 46 per cent from their farm, 40 per cent from trading and 10 per cent from other sources; 76 per cent of Hannah Akech's income came from remittances, 9 per cent from farming, 3 per cent from trading and 12 per cent from other sources. Between these extremes, there was no typical composition of income. There was also no clear relationship between the composition of income and size of income, with two exceptions. The two households with the highest incomes (the Owinys and the

Olals) also had the highest farm incomes and the households with the lowest incomes (Millicent Mboya, Jennifer Ogutu, Margaret Ogendo, all receiving incomes of under KShs.500 per month) had all-round low incomes.

Most of the households had a very small farm income: mean monthly farm income was KShs.247 and this falls to KShs.128 when the apparently anomalous Owinys and Olals are dropped from the calculation. Many of the inequalities in total income resulted from differences in the size of off-farm income, which came largely in the form of remittances and off-farm activities. Most of the best-off households received a large proportion of their incomes from off-farm sources and the poorest households were mostly those with comparatively small proportions of their incomes coming from remittances.[1]

The poorest households had low farm incomes and also received only small amounts of remittances. Some were widows whose children were artisans or clerks. These had large households of their own to support and had little money to spare for their mothers. Others were much younger women whose husbands were in the lower reaches of the labour market, moving in and out of insecure work. Other households (Angela Owete, Magdalina Osaro and, more dramatically, Benedicta Adhiambo, Helen Ombura and Hannah Akech) had comparably low farm incomes, but higher total incomes, because their children or husbands were in higher paid jobs and could send them remittances regularly. The Odhiambos' income in this period was raised by the receipt of KShs.3600 in bridewealth. Normally, their income was very low. Their children sometimes sent them a little money, but their off-farm income was low (see more detailed analysis below). Others (Rebecca Atieno and Rachel Akinyi) had low remittances and low farm incomes, but compensated by trading regularly. William Abuya had recently lost his job and returned to Koguta. He and his wife, Maretha, had an extremely low farm income, but William's brother helped him out. Maretha's mother also used to send her money. William also found some temporary work.

The households at the top of the income distribution, the Olals and the Owinys, were very different from each other. Peter Olal's household had a high farm income, but this in no way reflected its agricultural output. Peter's income came from his pension, rent from a shop he had built and occasional remittances from his children. He did not get enough money from these in the survey period and so he sold two cows to buy food. So his relatively high farm income actually reflects a loss of assets. The Owinys, on the other hand, had several children in secondary school, but no children old enough to send remittances. They earned almost all their income locally. Elizabeth traded in fish and other foods; her husband traded in maize in the hungry season. Elizabeth invested some of her profits in her farm, growing crops for sale. She was one of the few people in Koguta who sold crops on any scale. The household's total income from crop sales was KShs.11,757, compared with a mean of KShs.738 for the rest of the sample, and when the Olals are excluded, KShs.387.

Differences in the composition of households' expenditure were not so great as differences in income. Food accounted for more than 40 per cent of expenditure in twelve out of the eighteen households. With the exceptions of the Olals, who bought cattle, and Angela Owete, whose sons sent her money for ploughing, farm expenditure was less than 5 per cent of total expenditure. Spending on other consumption goods (such as non-food groceries and clothing) accounted for between 10 and 20 per cent of expenditure for most households. The contrasts between households were greatest in terms of expenditure on education, cash transfers to other households, and in the category of other expenditures (which included transport and health). Spending on education was quite closely related to income level. Better-off households were more likely to be able to pay school fees. The developmental cycle was also important here. Some of the households farther down the income distribution were in difficulty because the head was too old to be able to earn much money. These households were not usually supporting school-age children. Differences between households with regard to the transfers made to other households are analysed below (pp.152–6).

Understanding livelihoods involves working with two views of the household. One view is to see it as an economic unit, the other is to see it as the locus of relationships between individuals who are enmeshed in a web of relationships within it and beyond. Both views have to be kept in play. People's life chances and the choices they made were affected in many ways by processes that impinged on the household as a whole. Households' developmental cycles exerted a strong influence on people's ability to invest their own labour power in farming and on their access to the labour power of others. Developmental cycles also affected their ability to earn an income from off-farm activities.

Relationships within households also shaped livelihoods. Changes in intra-household relationships often centred on financial responsibilities, access to resources and income-earning strategies. The pressures of economic decline fostered new divisions of economic responsibilities and strategies for constructing livelihoods. In some households, the husband took on most financial responsibilities, made most of the important economic decisions and perhaps appropriated his wife's labour, for example for cash cropping. At the other extreme, the woman took on most financial responsibilities. She might have an independent source of income and make most of the important decisions. These variations between households were so great that they amounted to differences in household forms. A straightforward distinction between male and female-headed households could not hold, because male involvement in finance and decision making covered a spectrum from control to desertion. These new household forms were strongly related to differentiation in the bases of livelihoods, as well as to processes affected by household developmental cycles.

Households dependent on remittances

Some of the people who depended heavily on remittances were elderly widows who were getting support from their children. Their investment in their children's education had paid off. Sometimes their daughters-in-law helped with the farm work for a few years (Magdalina Osaro), but many women could not rely on getting this kind of assistance (see Chapter 8). Older women whose daughters-in-law now had their own fields and women whose sons' households lived in town had little access to labour. They might be too frail to do much work themselves (Angela Owete). It was remittances that allowed them to maintain a reasonable standard of living. Jennifer Ogutu's problems showed what happened when children could not give much support. Her income of KShs.200 was well below that of everyone else in the survey. Most of her very small income came from her farm (KShs.87). Her sons had inherited most of her husband's land, leaving her with a garden of half an acre, so she borrowed an acre of land in 1988. She did a great deal of farm work, despite being in her seventies. Her vulnerability would increase when she became too old to work on the farm regularly.

Other people getting remittances were women in their thirties and forties whose husbands, and sometimes children, were sending money back. The total amounts of remittances look quite high, but these do not give an accurate picture of the resources at their disposal. They all had school-age children and a lot of the money was allocated by the senders to pay school fees (Benedicta Adhiambo, Helen Ombura). They were also sent smaller amounts of money and these were vitally important for subsistence. For this reason, their remittances rose in the hungry season (*ndalo kech*) before the harvest. Women like this had often lived with their husbands in town for a decade or so before coming to live in Koguta for the first time. It was cheaper to bring up children in the rural area. School fees in rural schools were lower and produce from the farm made life cheaper. They also saw themselves as preparing the house and farm for their husbands' retirement. The survey therefore captured these women at the time in their lives when their remittance incomes were at their highest. The remittances would fall once their children left school and their husbands retired, particularly since the next generation would have even greater difficulties in finding work.

As their children grew older, women in this position had more time for trading. Because they belonged to a generation which inherited a subdivision of already small holdings, their farms would never provide a significant income. Benedicta Adhiambo's farm provided 15 per cent of her income (21 per cent if remittances for school fees are discounted from income). The equivalent figures for Helen Ombura are 16 per cent and 20 per cent. Wives of migrants often supplemented their incomes by weaving a few baskets each week, or, less often, by selling small amounts of crops or doing some very small-scale trading. They usually lacked the time and capital to trade more than this. Basket weaving was a popular way of earning cash because it fitted in so well with a woman's domestic work. Most women did not have the time to make more than about three baskets a week, which gave a gross return of about KShs.30.

There were several reasons why these women did not look for other ways to earn money. Their husbands were generally reluctant to give either permission or start-up capital.[2] There was also a powerful model of household responsibilities at work. Good husbands were supposed to support their wives. A woman who traded was making a public demonstration of the household's failure to conform to this ideal.

Households that could not depend on remittances: constructing a livelihood

Most people could not rely on a steady flow of remittances, so did not have the option of hiding their need for money. This much larger group of people was diverse. Some were women whose husbands were not sending much money back; others were widows and married couples whose children were not giving them much support. Others were younger married couples where the man had come back, at least temporarily. There were strong contrasts in the size and composition of income in these different kinds of household. A young woman whose husband was not supporting her would have more energy for trading or for intensifying production on her farm, but she would also be tied down by childcare responsibilities. If she needed money desperately she would have to juggle childcare and income-earning activities.

Margaret Ogendo was in this position. With a farm of only a quarter of an acre, and four young children, she needed to buy maize for eleven months of the year. Regular trading would have been too time consuming. She sometimes worked on other people's farms, as her co-wife was prepared to look after her children from time to time. Farm labouring usually took up only a few hours in the morning, and it brought in money straight away. It had a strong appeal for women who had no capital for trading and who needed money quickly. The drawback was that the demand for farm labour, never very great, peaked at just the time when a woman ought to be working her own farm. This period was also the height of the hungry season, when food prices were highest, and this put additional pressure on women in poor households to find cash. Crops were planted late and weeds were neglected, which lowered yields. A downward spiral could easily set in. Output from the farm would fall and the woman would need to work on other people's land more often. This is what seemed to be happening to Margaret.

A woman could expect to have about twenty years when she was strong enough to take on extra work and her children were old enough to be left. But this was the time when she needed money most, to pay school fees for children at secondary school. Katherine Okonya was moving into this period. Katherine's husband was working as a night watchman in the Rift Valley, but she had not seen him for two years and he sent her only KShs.100 over the year. Her two oldest daughters had just dropped out of school because she had not been able to pay their school fees. They worked on her farm and looked after the younger children while she did agricultural labouring and made baskets. Her relatives and friends also gave a lot of help (see the discussion of transfers on pp.152-6).

Even when they could earn money, people like these were in a precarious position. There were seasonalities in the returns to trading and the availability of farm labour as well as annual variations in farm output. Any of these could push a household into acute poverty. If an income-earner fell ill, the household might fall into chronic poverty. These vulnerabilities were becoming more intense and the numbers of vulnerable households were growing. The growing general impoverishment made it difficult to make much money from trading. It also reduced the demand for farm labour. Relatives and friends might help in a crisis, but this safety net was becoming more fragile.

What might happen when the husband returned was shown by the different experiences of the Mbogas and the Odhiambos.

The changing fortunes of two households

Judith Mboga and her husband, Charles, married in 1955. Until 1968, Charles had been a building worker, first in Nairobi and later in the Nandi Hills, in Western Kenya. He would send her money every month (she remembered a figure of KShs.30) and she used the money to buy maize. He also paid the children's school fees. Judith used to sell sweet potatoes from the garden in order to buy fish or meat, but she did not trade regularly. Joseph Odhiambo worked as a tea picker on an estate in Kericho from the time that he got married in 1954 until 1977. When he was at home, Joseph would trade in maize, transporting it with donkeys, which he owned. He was never very interested in farming and his wife, Martha, did almost all the farm work. Joseph's remittances to Martha were never large (she remembered figures of KShs.20–25 per month, which rose to KShs.50 in the hungry season) and she used some of the money to buy raw materials for trading in cooked food. Joseph also paid the school fees himself. So, while both these men managed to send small amounts of money home, their wives needed to supplement the money.

In 1968, Charles had an accident at work and lost his job. He had been at home ever since. This misfortune left the Mbogas very short of money. One possible solution was to grow crops for sale. When other returned migrants did this, they usually drew on household labour power and appropriated much of the crop income. The Mbogas could not follow this strategy, because Charles' injury stopped him from working on the farm. When the children reached secondary-school age, things became very difficult. In 1977, Judith decided to start trading more regularly.

> She began by making a lot of baskets and using the KShs.20 profit to buy *omena* (dried fish). Charles thought it was good. 'If I got, say, KShs.7 profit he did not take it from me. He knew I would buy maize with it. I traded in *omena* for two and a half years and my profits rose.' In 1979, she began to trade in second-hand clothes.[3] Charles gave her KShs.60.

She bought clothes from large-scale traders and re-sold them. 'It was my idea to do this. My brother was doing it.' After a year, she began to get the clothes from Kisumu. 'For example, I would get twelve shirts at eleven shillings each and I would sell them for about fifteen or eighteen shillings.' In 1982, after a clampdown on the trade in Kisumu, she began going to Nairobi. She did this every week. She now bought in bulk and sold to other traders.

Charles was happy about this. 'We started to drink tea, eat bread and make chapatis.' Her weekly turnover was KShs.3000 and she made a profit of KShs.400. She would 'eat' Khs.250 and keep KShs.150 for school fees. Most of the money was spent on school fees and food. The eldest son had trained as a primary-school teacher and three daughters were at secondary school.

Judith also used some of her trading profits to pay hired labourers to work 5.5 acres of the household's land. She sold tomatoes and onions from the farm. Both she and Charles sold some of the crops and kept the proceeds and Charles used some of this money to build up a herd of cattle. He had recently sold some cattle and used the proceeds to buy, jointly with Judith, a plot of land near the local secondary school, building on it a house to rent out to the school.

Judith's attempt to build a livelihood around trading seemed a resounding success. But it was fragile.

'I do have problems. The biggest one is the police. Once, they caught me in Kisumu and I had to give them a bribe of KShs.900. Another time, I had to pay them KShs.300. Recently, I sold a woman KShs.280-worth of clothes on credit. She refused to pay, saying that she couldn't be forced to pay, because the trade is illegal. Then she told the police and they came to the house at Sondu where I trade. I had to pay them a KShs.700 bribe. The other problem is that people buy clothes on credit and pay me back very slowly.'

As well as these problems, Judith was finding travelling to Nairobi every week exhausting. By the end of 1988, she was considering giving up the trade and selling new, locally made clothes instead, despite the fact that the returns would be lower. Then she broke a leg, putting all her trading in jeopardy.

After Joseph Odhiambo returned permanently to Koguta, the Odhiambos lost one of their main sources of cash, but Joseph continued to trade in maize for several years. He used some of the proceeds from his maize trading to make a long-term investment in cattle that could be used to pay the children's school fees. But he also spent quite a lot of money on alcohol and there was little money for food. So Martha decided to get a more regular income by working for a fish trader. She also sold small amounts of crops from the farm. By doing this, Martha earned enough to feed the household. Unlike Judith Mboga, Martha was not responsible for paying school fees, so she had neither such an incentive nor her husband's encouragement to undertake regular trading.

In the late 1980s, the Odhiambos' incomes fell sharply. The children had reached secondary-school age and Joseph had to sell the donkeys to pay school fees. Doing this stopped him from trading in maize. Then the cattle were stolen in 1988. At the same time, Martha, now in her fifties, had to give up working for the fish trader because she developed angina.

> 'I get money from selling chickens, vegetables and groundnuts and sweet potatoes. I'm the one who does this because I'm *wuon agulu*.[4] Why should I give Odhiambo money? Of course I don't give him any money. He doesn't use it wisely – he drinks. He gets money in his own ways. He doesn't give me any. My children give him less money than they give to me because he doesn't use it properly.'
>
> (Martha Odhiambo)

The Odhiambos had some temporary relief from their problems in the year that they took part in the budget survey. Joseph received KShs.3000 and Martha KShs.600 as bridewealth from their daughter's husband. Joseph spent KShs.1300 on an instalment of his own son's bridewealth, but he also bought two cows in order to start building up another herd. The long-term prospects looked bleak. Martha's health was sometimes too poor for her to work on the farm or make baskets. Although one daughter sent very small amounts of cash regularly, the sons sent nothing.

While Judith had the contacts needed to become involved in a far more profitable off-farm activity than any tried by Martha, she also had a far greater need to find one. Martha's husband took responsibility for paying school fees, while Judith's husband could not do so. By the time Joseph had sold his main source of livelihood, his donkeys, Martha's health was too poor for her to trade regularly. Bridewealth receipts temporarily relieved their slide into poverty, but it was not going to be staved off by remittances from the children. The Mbogas, on the other hand, while also threatened by ill health, had investments in cattle and real estate that would help to cushion them. These households show two of the many ways that gender relations, developmental cycles and differentiation could interact.

Conclusion: poverty and livelihoods

What stands out most clearly was how pervasive were the processes pushing people into greater poverty. All the major bases of livelihood were being undermined. Farm incomes were low and, for most households, were falling. Diminishing yields and the growing scarcity of land would make this worse. Only a few people could count on getting remittances regularly and these were under threat from the crisis in the urban labour market. Falling incomes meant that nobody was going to make much money from trading. Although some people managed to resist the effects of the erosion of one basis of livelihood by shifting to another, on the whole, downward mobility was cumulative. There were a few niches within this impoverished economy, as Judith Mboga's rising trading income showed, but her experience also demonstrates how fragile these niches could be; how vulnerable to the consequences of injury and disease.

Some kinds of household were especially vulnerable to impoverishment. In particular, if not compensated for by remittances, changes in cash needs and labour availability caused by the working out of household developmental cycles pushed a household further into poverty. But the position of most of these households was fragile; vulnerable to the vagaries of illness, insecure employment or risky trading activities.

People's ability to cope was shaped by other long-term changes in the rural economy. The most important were the impact of the shrinking agricultural resource base and people's varying ability to find alternative sources of income. Being able to make an adequate income from farming depended on access to land, labour and money. Long-term inequalities in the size of holdings reflected differences in the distribution of land when Koguta was settled, but there was also some appropriation of land through land disputes and land adjudication in the 1960s. Because of subdivision, there were also inter-generational differences in the size of holdings. These differences did not always feed through into inequalities in farm income, because many households with above average holdings lacked the labour and capital needed to make use of them. Most people had felt the effects of diminishing access to extra-household labour through non-market mechanisms and changing modes of household labour use (particularly the switch towards off-farm activities).

Differences in off-farm income were closely related to the availability and disposition of labour power within households. The ability of household members to engage in off-farm activities varied several times over the household developmental cycle. But women's involvement in trading, while largely the result of financial pressures, often hinged on the outcome of domestic conflicts over control of women's activities and incomes, as we shall see in Chapter 8.

A situation where the most significant resources came from outside the locality, and where many people lacked access to these resources, poses questions about relationships between households in a stark way. How did people react to this gap between people who are just managing and those who were not? How important were relationships beyond the household for getting access to resources? Did it make sense to invest in social relations? What obligations did people demand, and respond to? What did people think about all this? Did they have different perspectives?

Relationships beyond the household

The budget survey included questions about resource transfers between households. Several different patterns of behaviour emerged. In general, the kinds of transfers moving in and out of a household and its members' attitudes towards rights and obligations to their kin and neighbours depended on their socio-economic position. When the *dala* was still more unified, even though each *ot* was expected to grow its own food supply, redistribution was common. The senior woman in the *dala* redistributed the meat men brought back from hunting. Eating arrangements encouraged the transfer of resources, because the men and women ate separately, but communally, the food cooked by each woman in her *ot*. This arrangement gave plenty of scope for redistribution at the point of consumption. Each married male had his own field, *mondo*, which his wives cultivated. The grain from the *mondo* was used for beer brewing and sometimes converted into stock. However, its main functions were as a store for redistribution among the wives, evening out disparities. A man could also give grain from this granary to his cognatic kin without needing to consult his wives.

As labour exchange within the *dala* became less common, resources were redistributed less often. Shortages of land and labour made the *mondo* disappear in most homes, while the declining authority of the *wuon dala* removed the *dala*'s focal point. People began to eat separately in their own houses. Food sharing happened bilaterally between women who had a good relationship. Transfers between women from the same *dala* did not seem to be qualitatively different from transfers between women from different *mier*. Men and women did behave differently, however, and they need to be examined separately. Some patterns of transfers were reciprocal; in others the transfers mainly moved in one direction.

Reciprocal transfers ('borrowing') were common. Typically, a woman in a poor household would have a small group of other women (five to ten) she could 'borrow' small amounts of food from to cover shortfalls in her supply. They would do likewise to her. Women usually operated this form of insurance mechanism with close kin – other women in the *dala* and the *jokakwaro* – but they might also turn to friends, often from the same church. They might also occasionally share food with a few other women. Ruth Ogendo explained how she decided whether to start this kind of relationship:

> 'You choose by learning the person's character. If somebody came and begged for the first time, I would give and wait to see what happened when I myself begged from her. If she refused to give, I would also refuse if the woman came back a second time.'

Rispa Abiro agreed: 'You usually expect a return when you give.' Some gave and received more than others. This variation could be explained partly in terms of people's personalities and the state of relationships within a *dala* or a *jokak-waro*. Exchanges happened where there was *winjruok*. But women who could get resources in other ways were much less likely to 'borrow' and 'lend'. They might occasionally turn to others, but it would not have been rational to get heavily involved in these relationships. The fact that most women who exchanged food regularly found it difficult to maintain themselves meant that this channel of access to resources might easily fail. Although getting access to income through the market economy might also be fragile and uncertain, women who had the choice preferred it to the moral economy of 'borrowing'. For the poorest women, who did not have many alternative sources of income, their access to food and cash through social relationships was vital to their survival. Because men in poorer households usually did not take responsibility for the food supply, men did not usually give and receive food. They were more likely to 'borrow' small sums of money from other men.

This kind of borrowing and lending was very small scale. Cumulatively it could make the difference between managing on a day-to-day level and going seriously short of food, but people could not rely on kin and neighbours to lend on any scale. Apart from bridewealth, transfers of larger amounts of food, cash and clothing tend to come from a woman's husband, or other male affines and cognates and, for men, their male agnates. Women could rarely give much cash to their kin, but they might receive cash from them. They might get help from their mothers, brothers or other members of their natal *dala* in a crisis, or they might get regular remittances. Money coming from kin explained much of the differentiation between households. The most economically secure households received regular cash transfers from spouses and children. Many others were protected from a fall into extreme poverty only by a rallying round of their cognatic and affinal kin.

The most vulnerable households were a heterogeneous group. They included elderly widows who could not rely on anyone for regular support and much younger women with children whose husbands were not supporting them. What these people had in common was that they did not have a reliable channel of access to the remittances coming into Koguta from outside. They were often short of land or labour, meaning that they grew only enough grain to last for a month or so. They relied heavily on transfers, regularly giving and receiving food. People like this found it much harder to reciprocate larger transfers from the better off. The only way to get resources from them was to use the language of kinship as a claim, a demonstration of self-evident obligation.

It sometimes worked. Elizabeth Owiny, a prosperous trader, gave Katherine Okonya four blouses for her children in June 1988. Katherine explained that 'she is an elder "co-wife" from the *jokakwaro*, so she should help people'. Better-off people who subscribed to this ethos used the same language to explain their transfers. 'I call her mother-in-law/co-wife/daughter-in-law' was offered as a sufficient explanation. To recognise such a relationship was to accept rights and obligations. Because they did so, the record of transfers from some better-off households was extremely one-sided. In June 1988, Elizabeth gave out a basket of maize, some sorghum and KShs.5 to the assistant deaconess of her church 'because she is a widow'. She gave some fish and cooking oil to a close woman friend in a neighbouring *jokakwaro*, 'because she calls me daughter-in-law' and some fish and paraffin to an old woman in her *jokakwaro* 'because she has no husband and children and is very old'. She gave a sweater and dress to her mother, 2 kilograms of maize to her mother-in-law and 4 kilograms to a woman from the *jokakwaro* whom she calls 'co-wife'. She gave maize flour to another 'co-wife' 'because she has problems'. To another 'co-wife' she gave soap, paraffin and 'other small things'. She gave her sister KShs.2 and another KShs.2 as a funeral donation. She gave a 'sister' from the *jokakwaro* a headscarf and some soap. She gave 'small things' to other people, but she cannot remember them. 'These people come very often.' They didn't always beg (*kwayo*); sometimes she just gave them something ('*amiyo amiya*'). In the same month, all that Elizabeth was given was a handful of groundnuts from a 'co-wife' in her *jokakwaro*. She commented: 'People don't give much to me because they think I have enough, because I give. I have tried to beg, but they say I'm laughing at them when I do that and so they don't give to me.'

Elizabeth's transfers-in came less often, in the form of cash from her migrant son and daughter-in-law. Contrast Elizabeth's record of transfers-out for June 1988 with the following, for June and August 1988:

Peter Olal and John Odhiambo were two of the most prosperous men in Koguta. Both were getting pensions and, as we have seen, Peter rented out a house. The Olals had, in effect, withdrawn from the network of transfer-relationships linking other households. John Odhiambo's transfers were of a different order, part of a strategy to enhance his social role as a male elder. A newly retired migrant, he was on the board of governors of several local schools

Table 7.1 Transfers out of two Koguta households

a) Peter Olal		*Helen Olal (Peter's wife)*	
June 1988		*June 1988*	
Recipient	*Transfer*	*Recipient*	*Transfer*
–	nil	woman from *jokakwaro*	2kg maize and vegetables
August 1988		*August 1988*	
Recipient	*Transfer*	*Recipient*	*Transfer*
–	nil	–	nil

b) John Odhiambo	
June 1988	
Recipient	*Transfer (KShs)*
Step-mother's daughter	100 Loan – to be taken out of her wages. She works as a domestic servant for John's daughter
Friends	100 Loan
	200
	200
August 1988	
Recipient	*Transfer*
Funeral (daughter's husband's grandmother)	240
Harambee	200
Harambee	100
Harambee	15
Man from Rarieda	400 (loan repayment)

(hence the *harambee* donations).[5] The funeral contribution was the size expected from a senior kinsman. He moved in quite a different social field from his neighbours and his loans were almost all made to men outside his lineage whom he knew through the church or his business activities.

In a world where most people are very poor, people with slightly more resources had to make a choice. If they recognised the claims their kin and neighbours made on them, they would be locked into redistributing their incomes. The only way to withdraw from the web of redistribution was to withdraw socially and look elsewhere for one's own networks. Mrs Odhiambo commented, 'Our friends come to visit us in cars.' The most effective way to withdraw was to live somewhere else – to stay in Nairobi or buy a farm in the Nyanza Sugar Belt, well away from the claims of poor relatives.[6] People who could not afford to live somewhere else had to use different tactics. Some discouraged casual callers. Others joked about *jowanya* ('people who turn up at meal times') and tried to keep their food stocks out of sight. Poor people felt that they had lost something important:

> 'In our *jokakwaro*, we can ask (the richer members) for food once or twice, but after that they will refuse and say, "Prepare your own". Your friends will help you more than the (richer members of the) *jokakwaro*, with money, food and clothes. It's easier to ask friends than to ask the *jokakwaro*. *Jokakwaro* will gather together (to discuss what to do about your problems), but your friends will help you straightaway. These days, people turn more to friends than to the *jokakwaro*.'
>
> (Rispa Achieng)

Notes

1   A large proportion of the inequalities in remittances received are related to differences in the labour market position of migrant husbands and offspring. This finding contrasts with Collier and Lal's suggestion that remittances are an equalising component of rural incomes (Collier and Lal 1986: 264–6).
2   Gender ideologies are discussed in Chapter 8.
3   The trade in second-hand clothes, though illegal, was one of the most lucrative forms of trading. Much of the clothing originated in the USA. People in Western Kenya believed that it had originally been donated to charities and that somebody was making a lot of money out of selling it wholesale. The clothing was popular because it was thought to be better made than new clothes produced locally.
4   Lit. 'The owner of the cooking pot'.
5   A *harambee* is a common method of fund-raising in Kenya. Donations are made to fund community projects, such as schools, with the collection usually being made in public. Making plenty of *harambee* donations is an important way of gaining social prestige.
6   Cf. Cohen and Odhiambo 1989: 54:

> The sugar belt lands were quickly seen as a safe haven for accumulations produced by better-paid occupations and professions ... the rural world ... offered no opportunities for forward-looking Luo, indeed the countryside had an impoverishing effect with its limited production and immense claims upon savings ...

8 Gender and livelihoods in Koguta

The upheavals the people of Western Kenya have experienced since the British first arrived – the rise of labour migrancy, the growing importance of money in all aspects of daily life, the encounter with mission Christianity, Western education and urban life – have changed the fabric of relations between women and men. They share this experience with most people in rural Africa. Cash cropping precipitated and, often, rested on changes in gendered divisions of rights over land, labour and products and in the nature of the domestic unit (Wright 1983; Guyer 1997; Mackenzie 1998). Chapter 4 showed how commercial farming weakened women's power and authority. The emergence of migrant labour economies rested on reorganisations of household forms, divisions of labour and domestic responsibilities, as did later economic changes in these regions (Murray 1981; Hay 1982; Bozzoli 1983; Sharp and Spiegel 1990; Moore and Vaughan 1994).

In Koguta, the house, *ot*, and the group of people within it, *jokamiyo*, grew in economic importance at the expense of the *dala*, and the authority of the *wuon dala* over his sons and their wives was weakened. That is not to say that *udi* simply became isolated units. *Udi* in the same *dala* were linked to one another in subtle ways. There was plenty of scope for tensions and jealousies – if one son had found a better job than the others, or if he had been given more land – but if the women were on good terms, they might borrow food from one another, help each other out with weeding or keep an eye on each other's children. A *jadala* with a job in town might help pay school fees for a sibling's children or help them find work. *Jodala* had a sense of a common family history and values. However, it was the *ot* that mattered for the basics – farming the land, bringing up the children and the daily challenge of making a living. 'The household' was not an abstraction imposed by social scientists, but the core social group for constructing livelihoods. Even so, it might well not act as a unit. One of the main issues for households in Koguta was whether they *were* units, whether their members did have common interests, whether they were bound by rights and obligations and what these were. For most people, getting the resources they needed to make a living depended on their membership of a household. Women could get access to land only through marriage. A migrant man who was concerned about keeping a base in Koguta needed someone to look after his interests at home. The terms on which these interdependencies were worked out were not a script, but a terrain for discussion, or contestation.

Households were constructed on a resource base that was shrinking. Most young women marrying in would never get more than a fraction of an acre of land. Most young men would find it very difficult to earn a wage that was large enough to support a wife and children. The interdependencies that bound members of a household together were becoming less obvious. In some households, people were fending for themselves, rather than co-operating in a joint strategy: the household was fragmenting.

A simple model in which the spread of markets undermined women's powers would be a quite misleading way to understand the links between gender and agrarian change in Koguta. Changes in gender relations were slow and piecemeal and they did not move in a linear way. Instead, they responded to a series of changes in the political economy of the region – the emergence of labour migrancy, the rise of the remittance economy and more recent problems in the urban labour market. Nor did gender relations change uniformly. They looked different in households with different types of livelihood. This was because, as we have seen, different livelihoods create different pressures and raise different issues within households. Finally, while gender and livelihoods have been closely linked, gender relations could not be read-off from livelihoods. As well as power relations, ideas and arguments about rights and responsibilities have been critically important.[1]

'His job was his *mondo*': labour migrancy and gender

The British arrived in Nyanza at a time when the local economy was changing. As people moved from the drier areas around Lake Victoria into higher, better-watered ground, they were shifting from a way of life based mainly on pastoralism towards greater reliance on agriculture. The rinderpest epizootic of the 1890s, which temporarily devastated the herds, intensified this shift towards agriculture. There must have been great local variation in the relative importance of pastoralism and agriculture. Accordingly, there were local variations in the sexual division of labour. Nevertheless, there were some common features. Young men herded cattle, hunted and went on cattle raids. Women did most of the agricultural and domestic work. In richer homes, adult men's work in agriculture was limited to bush clearing. Poorer men also hoed, weeded and harvested the crops. Agriculture was remembered as being the domain of women, and secondary in importance and status to pastoralism (Whisson 1964; Hay 1972; Pala 1977; Butterman 1979; oral evidence). Colonial officials, and Luo people themselves, drew on this stereotype to explain why there is so little commercial farming in Luoland.

Married women were responsible for feeding their own *jokamiyo* from the fields they were allocated. JoKoguta who grew up before the 1930s agreed that a married woman was *wuon puodho* (the person in charge of the field). She could decide what to plant in her own fields and she had control over the crops coming from these fields, including any surplus. Her husband was *wuon lowo*, with ultimate authority over the land.[2] A husband also had his own field, *mondo*. His wives did most of the work on the *mondo*, and he was expected to use the crops to make good any shortfall in their supplies of food. Otherwise, he could dispose of crops from the *mondo* as he saw fit. A man could accumulate power and prestige by exchanging surplus grain for cattle, which could be used as bridewealth. Men's ability to accumulate resources rested on control over women's labour (cf. Guy 1990).

Before the 1930s, most migrants were unmarried men, who played a minor role in agriculture, or they were squatters, who tended to migrate as whole households. During the 1930s, many of these early migrants continued to work outside after they got married and their labour began to be missed. There were many jobs on the farm and around the home that could not wait for their return. Most importantly, somebody had to hold the rural household together and protect and manage the land. Margaret Jean Hay distinguishes between women's roles as owners, occupants and managers of property. Women in colonial Western Kenya shouldered much of the burden of household reproduction at the same time that economic decline eliminated the grain surpluses that had been one of their few avenues to control over resources and property accumulation. Instead, women took on the roles of occupants of household land, validating and protecting men's rights in land, and *de facto* farm managers. Women in poorer areas became placeholders (Hay 1982; see also Hay 1976). The kinds of responsibilities that women took on made it unlikely that labour migration would lead to agricultural growth. They became guardians of the land, rather than managers of a farm enterprise. This guardianship strictly limited their control over resources. A woman should guard the land and livestock, but she would not be able to acquire or dispose of these resources. Nor would she have the money needed to intensify farm production. It was a recipe for stagnation.

As we shall see, this picture did not apply to all rural households. Nor was it static. Power relationships within households responded first to the pressures of labour migrancy and later to the reductions in remittances that accompanied problems in the urban economy after the early 1970s.

1930s–1950s

Until real wages in the urban economy began to rise in the 1950s, most Luo migrants did low-paid work, first mainly on the European farms and plantations and later in the urban areas. However, as we saw in Chapter 6, a few men found better-paid work as teachers, clerks and other jobs in administration. Inequalities in wages did not bring about large differences in productivity between different farms, however. Nor was there much difference in the roles of women in different migrant households. Most migrants' wives stayed at home, growing crops to feed their children, perhaps doing some seasonal trading to earn some cash. Men were too far away, for too long, to play much of a role in day-to-day household decision-making. But there was no evidence that wives of better-off migrants were able to use remittances to expand farm production. The suggestion that some women were able to do so is central to Kitching's hypothesis about a dynamic link between off-farm and farm inequalities in the smallholder areas of Kenya (Kitching 1980). Migrants had other priorities. As we have seen, few men invested in farming beyond what was needed to replace their own labour power. The first migrants spent their wages on taxes, clothing and bridewealth. Migrants in better-paid work also made other investments. They bought cattle and paid school fees. Most migrants would not have been able to afford to use their wages to expand farm production, but better-paid migrants were also reluctant to do so. Low prices, poor infrastructure and distance from markets limited the returns to farm investment. But the existence of men like James Onditi suggests that these were not overriding problems. Few migrants were willing to delegate financial responsibility and decision-making power to their wives. This reluctance stemmed from a deep-seated distrust of women's reliability that is explored in more detail later. Nor could migrants easily ask their male kin who lived in Koguta to supervise their investments. The growing importance of the conjugal unit at the expense of relationships beyond the household made it difficult for relatives to intervene. Men's reluctance to send money to their wives for them to buy farm inputs could probably also be explained by their concern to put their earnings into investments, like cattle, that could not easily be converted into money when food was short (cf. Ferguson 1992). For migrants who thought they might be able to improve their position in life, money from outside became more important than farming. In the past, it was men who made investments. They acquired cattle and used them for bridewealth. More wives meant more land, more granaries and more children. Now that it was money and education that meant the difference between poverty and moving up in the world, men dominated access to these as well. So, despite the differences in wages earned by migrants, divisions of labour between migrant husbands and their wives were fairly uniform. Men's reluctance to delegate financial responsibility to their wives was one of the factors that weakened the impact of differences in wage income on the rural economy and slowed down the growth of inequality.

In this period, there was much more change in the division of tasks and spheres of authority and responsibility in households where the husband was living at home. For a few decades, some men had enough land and livestock to make a living locally and were able to avoid having to look for work. A few were even able to accumulate resources through farming and trading. This often involved bitter conflicts over access to their wives' labour and over the question of who should control income from the farm. Men who relied on farming usually took charge of the operation by appropriating their labour and keeping the crop income. As we saw in Chapter 5, Peter Odiyo's single-minded concentration on farming enabled him to generate enough of a surplus to give his sons secondary education. Speaking about the 1940s, his first wife talked about the conflict this strategy involved.

'Odiyo is a fierce man and he liked to quarrel. The arguments happened because he expected me to do a lot of work, and I didn't like quarrelling. I did think about leaving him, but my bridewealth was high, so I couldn't leave. He used to threaten to bring another wife. I would say, "Go ahead", because I was fed up with all the arguments. He married two other women. After they arrived, I was left with one field and I worked very hard in it. The cattle Odiyo used to marry my co-wives came from my work.

These days, I don't work on the *mondo*. You only work on a crop that will help you. Odiyo is mean with money. He doesn't give any to his wives. He puts it to work; he buys cattle and keeps it to educate the children.'

Martha Odiyo

Here, the bargain was explicit. A husband would be more likely to get his wife to work for him if he gave her money and food. Martha stopped working on Peter's *mondo* once her children left school and he stopped giving her money for school fees and clothes. Instead, she got money from selling crops from her field and from her children.

Growing land shortage meant that the field system became simplified. Men who wanted to have a separate crop income, and who had enough land, continued to keep the *mondo* separate from other fields. Where there was less land, households began to amalgamate contiguous land into one field (or one field for each wife, if there were several wives). These farms might look similar – a single field – but the status of the field was an important issue. Was it all the *mondo*, or had the *mondo* disappeared? These were important issues, because they were also questions about rights to labour and crop income. Recognising someone's claim to be the *wuon puodho* involved recognising a moral claim to make decisions about the use of labour on the land and to keep the income from crops that were sold. In some households, the fields were amalgamated under the authority of the husband – in effect the whole farm became the *mondo*. In others, the husband took little interest in farming and the wife ran the farm: she was *wuon puodho* over all the farmland.

Some men living in Koguta, particularly those who were older, continued to have a *mondo*. Margaret Ojwang, widow of Philip Ojwang, married him in about 1940 and was the youngest of his four wives:

'He would plough and then he would order us to start planting. He would tell us to plant a certain crop, like groundnuts, and he would walk around the *mondo* and tell us what work to do. The wives did all the work in the *mondo*. He would just walk around telling us what to do.'

Other men had too little land for this arrangement. Jennifer Ogutu was born around 1906. Her husband lived at home and worked as a policeman. He did not have a *mondo* 'because there was very little land'. He did not work in her field and did not tell her what to do in it.

Married migrants could not easily farm a separate *mondo*. It was too difficult to time their return to fit in with periods of demand for labour on the farm. They also found it hard to supervise their wives' labour from a distance. Rather than starting a *mondo*, they allocated the land among their wives. When these men returned, whether it was after a brief spell outside, or whether it was after many years, they often could not start a *mondo*. There was no suitable land that was not either already under crops or needed for pasture. Other men who did have enough land to make starting a *mondo* an option chose to put their energies into cattle herding, or into off-farm sources of income. Millicent Mboya, whose husband had a job locally, explained, 'My husband did not have a *mondo*: his job was his *mondo*'.

In summary, changes in the field system called into question norms of entitlement to control over decision making and access to labour and crops. Before mass labour migration, being *wuon puodho* over her field endowed a woman with the responsibility for finding the bulk of her household's food supply, as well as the right to make planting decisions for that field and disposal rights over any surplus produce for it. Where there was no separate *mondo*, the man could claim that he was *wuon puodho* over all the cultivated land, and, hence, had the right to make planting decisions, appropriate his wives' labour and keep the crop income. In some households with a small *mondo*, the husband made similar claims over his wives' fields. Where the husband or his wives wanted to sell crops, this claim also covered crop income. Such a claim by a husband clashed with a woman's claim to the right to control her household's food supplies.

The outcomes of conflicts over these competing claims varied. The norm that each wife was responsible for her own food supply and had the concomitant right of access to the means of production militated against the unification of farm land in polygynous homes. It became the most common arrangement in monogamous households, although a few men still had a *mondo* in the late 1980s. Male control of land and crop income became most common in households where farming was the main source of income for the husband. But this was not always the case. Some women successfully claimed a share in the fruits of their labour.

Peter Olal, who was employed by the Agriculture Department locally as a veterinary assistant from 1942 until he retired in 1973, grew crops for sale, but hired labour to work on the crops, rather than ask his wives to work on them. His explanation for this arrangement was that they would otherwise want to take the crops to market and keep the money: a person who does the bulk of the work on a crop should get the money when it is sold. So, where a husband could afford to hire in labour, conflict could be avoided (cf. von Bülow and Sørensen 1988). Cases like this were rare, however. Most men who wanted to intensify production could not afford to hire labour and defuse the conflict. Unless they could win the argument, their wives lost control over the labour process and the crop income. At the same time, their husbands took on greater responsibility for household reproduction. Women tended to keep control over the labour process and the disposal of crops when their husbands were mainly involved in cattle herding or in off-farm activities and so were less keen to take over farm decision making. In these respects conflicts in non-migrant households parallel later conflicts in the households of migrants.

It was hard for a woman in this position to do more than grow crops to feed her household. Many turned to trading, which further reduced labour input into the farm:

In the 1940s Jennifer Ogutu would buy millet in Karachuonyo (South Nyanza) and take it to Ahero (in Kisumu District) on foot. 'Many women did this. We traded because the men did not find enough work.' She then started selling maize. Along with many other women, she would fetch maize from Kipsigisland and sell it at Sondu market. They would also fetch fish from the Lake and sell it at Sondu.

Changing spheres of authority and responsibility: the erosion of women's authority and new divisions of labour

Chapters 5 and 6 showed how farm output fell after the Second World War. People now needed cash to buy the food they needed to live. Alongside this, new household goods – clothing, furniture, cooking pots – appeared in the markets of Western Kenya. People began to put sugar in their porridge, to drink tea and wash with soap. More and more of the essentials of daily life could only be found in the market. The postal order from Nairobi was what made the difference between being comfortable, a *janyalo*, and living in poverty. The reduced opportunities in the urban economy from the 1970s onwards meant that fewer JoKoguta could rely on remittances to make a living. *Jonyalo* were the exception.

The transformation of the basis of the household economy from the farm to cash income changed expectations about who was responsible for what in the household. It also intensified conflicts over the authority that this responsibility should confer. Who was responsible for making sure that people had enough to eat? Could a woman demand money from her husband to buy food, or was it her duty to find the money? Were clothing and school fees the man's responsibility? If so, did this mean that he should decide how much to spend, and when?

In a dwindling number of households, the growing need for cash was met largely from remittances. Some of these households consisted of the wives and children of migrant men doing clerical or supervisory work. Others contained men who had retired from this kind of work with a pension. In both types of household, the bulk of remittances tended to come from a single source and be received by one household member, the woman in the former case and the man in the latter. Either arrangement left the wife almost entirely financially dependent upon her husband and responsible for only the details of budgetary management. There was less pressure on her than there was on other women to engage in independent economic activity and they tended to play a very passive role in their household's financial affairs, simply receiving a regular allowance to cover food purchases. Household reproduction was fundamentally dependent on male earnings and women's role in reproduction was limited to providing domestic labour and a security function through their farm production.

This model of the household as an economic unit, with a male breadwinner and dependent wife, was the one people reached for when asked about domestic divisions of responsibility and power. A 'good husband', *dichwo maber*, gave his wife enough money for her to feed the household, as well as providing clothing, household goods and school fees. This ideal was quite different from gender relations in pre-colonial Kenya. Most societies in what later became Kenya were patrilineal and polygynous, with domestic relations resting on a house-property complex, as was a broad band of groups stretching from Sudan to South Africa (Gluckman 1950). Under a house-property complex, property was divided between the 'houses' formed by a man's wives. Oboler has argued that many ethnographic accounts of house-property complexes in Eastern and Southern Africa have overlooked the rights over cattle they gave to women (Oboler 1994). Similarly, accounts of pre-colonial gender relations which I collected from informants in Koguta suggested that women had considerable authority over the fields allocated to them by their husbands. This was linked to the responsibility each married woman bore for feeding her own house (cf. Hay 1972, 1976, 1982). Missionaries preached the virtues of monogamous marriage, and a division of labour between male providers and female homemakers (Gaitskell 1990; Bowie *et al.* 1993). This preaching was reinforced by explicit state ideology, as well as by the encounter with contemporary Western models of the nuclear family. All this was reinforced by the fact that avoiding poverty depended so greatly on a resource – paid employment – to which men dominated access. People who described themselves as 'modern' and 'educated' prided themselves on this arrangement, which sat uneasily with other self-consciously 'modern' ideals of gender equality. The 'breadwinner' model gave women a claim on men's resources, and men a justification for male control and female dependence. Both claims were threatened by the pressures undermining men's ability to play breadwinner.

In households where men were the main source of money, women's growing financial dependence on their husbands meshed with these new models of domestic relations. Together, they modified the ideal of a good husband and proper division of responsibilities. In the years before the beginning of mass migrancy, women expected to have authority in their own spheres – to decide what to plant in their fields, how much time to spend working in them, whether to exchange some of the crop for sheep or goats, in other words, to be *wuon puodho* and *wuon agulu* (the person in charge of the cooking pot). When the cooking pot needed money to fill it, women became more dependent on their husbands. This dependence engendered a more unified model of the household, one that gave men ultimate responsibility for feeding their families and the authority that came with this responsibility. Trading carried a stigma of poverty and women who could avoid it prided themselves on not having to 'sit in the market'.

Claims about responsibilities are claims about authority, but they are also claims about duties. To acknowledge a man's authority and his financial responsibility was to demand that he fulfil his responsibilities. Recognising that statements about who had *teko* (the power to decide) over what involved making claims about duties, as well as authority, problematises the descriptions people gave. My questions about *teko* could be taken to be an enquiry about what usually happened in the homes of JoKoguta, what should be happening in that particular home, or what was actually happening. The budget survey helped me to navigate through these uncertainties, by providing information on decisions about specific flows of money as they were made. These could be compared with more stereotyped descriptions given at other times.

However, the stereotypes were also revealing, because they showed discontinuities as well as continuities with (no doubt equally stereotyped) descriptions of the past. Women, it was said, continued to have *teko* over matters to do with cooking and bringing up the children. Men still had *teko* over the acquisition and disposal of land and stock, over building and repairing the house, over children's marriages and bridewealth, over the organisation of funerals and over women's travel. They were attributed *teko* over the new issues posed by the cash economy. These included decisions about children's schooling or other major outgoings like furniture or medical fees. Male *teko* in farming, at least 'where the husband is interested in farming' (Millicent Mboya), also became the norm. In the stereotypical pattern, then, men took on new powers of decision making in a more unified household.

A modern marriage?

John and Margaret Odhiambo had only recently come back to live in Koguta. John had been a manager in Nairobi and had taken early retirement. Margaret had once been a nurse, but had given up work to look after the children. The Odhiambos had built a big, suburban-style house next to John's father's *dala*. They bought some cows and started to plant crops. They were keen to stress how they lived a life apart from their kin and neighbours.[3] They were 'educated' and this made their marriage different from other people's, yet also an example that others might move towards:

'How do women look at marriage?'
'Well, you have to talk about educated women and uneducated women separately. Uneducated women stick more to men.'
'You mean they're more dependent?'
'Under a man's thumb, yes. Without his agreement, a wife can't do anything, but that's also true of educated men. Of course, some women are good at manipulating and they may really have more authority in the home than the man, but they still pretend to be submissive. Educated couples spend more time together. Uneducated women tend to spend more time with other women. They live in different worlds from men. People comment on how much time we spend together ... '
'Are ideas about responsibility changing?'
'Yes, because of Christianity and education. Men now have more respect for women. They sit and talk. Uneducated men also have a little more respect, because of Christianity. There is more co-operation than there used to be.'

The Odhiambos were self-consciously 'modern'. Modernity involved a different kind of marriage from the norm, a more private world where the conjugal bond was more important than ties with kin beyond the household. The Odhiambos stressed that they took decisions together and that this marked them off from their less educated neighbours.

The budget survey showed that John received most of the income.[4] In June, his total income was KShs.24,000. KShs.21,000 of this came from selling his car. He judged that he would not need one anymore. It had been useful when he lived in Nairobi, but now he wanted to buy cattle. He spent KShs.11,000 on cattle that month. The rest of his income came from his monthly pension and from selling milk. In the same month, Margaret's income was KShs.770. She earned KShs.200 of this in her part-time job as an infants' school teacher. Most of the rest came from John and she used it to buy food. In August, his income was KShs.3400. KShs.2600 came from his pension. Most of the rest came in loans that he used for a funeral donation (see Chapter 7, p.155). That month, he did almost all the shopping. He said that he often did this. Margaret's income was KShs.528. It came from her job, from selling milk and vegetables and a gift from her son-in-law. She spent little of it. So John paid for most of the food, either directly, or by giving the money to Margaret. She was not completely financially dependent on John, but he was taking responsibility for the food supply.

What this meant for their relationship became clearer when Margaret fell seriously ill with malaria. John offered to take her to the hospital in Nyabondo and told her to go on ahead. She went, and got some treatment, but he did not appear. The doctor wanted to admit her, but she felt that she could not go in without John. Only he could afford to pay the fees. She struggled back the three miles to the house. John refused to let her be admitted, saying that the drugs she had been given would be enough. I found her lying on her bed, bitterly complaining, while John sat outside the house:

> 'He only cared about me when I was well enough to work like a slave for him. When I am ill, he doesn't care. Then I am just a body. Next time I'm ill, I'll go straight to Nairobi, where there are people who will look after me.'

Margaret was more isolated than other women. Her daughter-in-law was in Nairobi and she didn't have much time for the other women in the *dala*. She often complained that she was lonely. Nevertheless, it was the women of the *dala* who rallied around and took care of her when she was ill.

So even the Odhiambos, the exemplars of the unitary model, did not really conform to it. Whether the strategies of household members did conform to common goals was an issue lying just below the surface of discourse in Koguta. Women complained about husbands who did not support them and their children. Parents complained about uncaring children.

> 'My children stopped helping me somewhere between one and two years ago and my sons haven't visited for a year and a half. My daughter is working. She sometimes sends money, when she remembers, but my sons don't send anything. … It's not because they don't have enough money, it's because they don't want to. One of my sons is a driver, so he gets a good wage. Children in town despise their parents. They want to drink and smoke and they have their wives. Nowadays, children don't love their parents.'
>
> (Martha Odhiambo)

The ideal of an economically unified household was a claim on men's, and children's, resources, by other household members.

But the ideal could also justify restricting women's access to money. Many men felt ambivalent about their wives getting involved in any independent economic activity. While many women were turning to off-farm activities, men who were just about able to sustain their households' cash needs preferred not to give their wives scope for gaining an independent income. Although the wife of a migrant almost inevitably played a greater managerial role in the household than the wives of most non-migrants, a greater decision-making role for women was considered to pose a threat to the authority of men. This threatened the unity of the household and ultimately the whole social order (cf. Parkin 1978). It was a topic men, and women, quickly warmed to, using a discourse that combined fear of the supposedly chaotic potential of women's sexuality with a stress on the divided loyalties of a woman who has always married in from another clan.

'Luo men do not like their wives to trade, because they think that they will walk around.'

(Judith Achieng)

'Women don't know how to look after money properly. They take the money, leave the house and wander around the country – to Mombasa, Nairobi – and they want to show off, for example, by buying clothes, but a man will want to use the money properly, for example, by paying school fees … .

'Very few Luo women trade because they are weak (compared with Kikuyu women). Their husbands are not willing for them to trade, because of what they might do when they are away from home, like going with other men. Some women are trusted and they do trade like this, but it is better for the woman to be older.'

(Philip Ojwang)

'Luo men take their wives' trading income … a Luo woman who travels around trading is called a prostitute.'

(Mary Anyango)

'There are not many ways for a woman to get money around here. So when a man sees his wife with money, he gets suspicious.'

(John Obala)

These comments convey a basic distrust of women – their loyalties and their sexuality. A woman was an outsider who had married in from another clan and she could rarely be completely trusted. She had a rampant sexuality that needed to be closely controlled. The best way to do this was to restrict her movements. A mobile woman was suspect, because movement and sexual laxity were practically synonymous[5] and both threatened the cohesiveness of her marital home. A woman with her own resources might leave her husband for another man. But as well as being based on fears of women's latent sexuality and unreliability, the wish to control the movement of women had its origins in the reorganisation of the sexual division of labour on which labour migration depended. For the rural household to be maintained, it was essential that someone remain to hold it together. Regular trading, with the continual travelling it required, threatened this pivotal role. The disquiet felt by many men at the thought of allowing women to get economic resources independently also stemmed from a fear of power struggles between a man and a 'strong woman' (*dhako ratego*). Relationships within households were sometimes described as a zero-sum game:

'When the man is up, the woman is down and when the man is down, the woman is up.'

(Margaret Ojwang)

And the key to who was up or down was relative economic power.

> 'It is income that makes a woman strong. The thing that makes a man strong is that he does not want to be conquered in terms of getting income for the family.'
>
> (Peter Olal)

As in any collection of beliefs, Luo gender ideologies are not all congruent and people voiced contrasting attitudes to domestic relationships in different contexts. When describing ideal domestic relationships, both sexes drew on the concept of *winjruok*. In a home where there was *winjruok*, men and women co-operated according to their expected roles. In the case of men this amounted to using their authority benignly. People often used the word 'co-ordination' when they spoke about this ideal in English. It could be achieved only if women accepted the authority of men. An overly assertive woman (*jalelo*) could undermine it.

Women's domestic power had been weakened in other ways. Older women had less authority over younger women in the *dala* than had been the case a generation before and they also had much less access to younger women's labour. Mothers-in-law and senior wives still expected respect from their juniors, but they described their relationships with their own seniors in very different terms from the ways in which younger women saw their relationships with them. Making this comparison showed changes, as well as some continuities in relationships between different generations of women.

First the continuities. When a young woman first came to live in her husband's home, her status was ambiguous, because the early stages of marriage were seen as provisional. Because bridewealth was paid in instalments, either side could for a long time end the marriage fairly easily. A young woman's worth was unproven until she had given birth to several children. The ambiguity is reflected in the fact that a new bride (*miaha*) was also referred to as a guest (*wendo*).[6] A *wendo* would not be given her own field straight away. She would spend the first few years of marriage proving her worth as a wife, and living in the house of a senior woman. This woman might be her husband's mother. If the young woman was the third or later wife (*rero*) of an older man, she was supposed to live with a senior wife as her *nyar ot* (lit. 'daughter of the house'). The third wife was supposed to be *nyar ot* to the first wife (*mikayi*); the fourth to the second (*nyachira*) and so on. *Nyachira* could not be *nyar ot* to *mikayi*, because they were often closer in age, potential rivals. It was also likely that the husband's mother would still be alive when *nyachira* arrived, so she would be attached to her. The *nyar ot* relationship had become rare among younger women, because very few men married more than two wives.

While she lived with a senior woman, the *miaha* would cook with her, work with her in her field and do many other domestic jobs like fetching water and firewood. This was the period in which her contribution of labour to the older woman was heaviest. Before most young men began to undertake long-term labour migration, and for a long time afterwards (until young wives began to spend the first years of marriage with their husbands), young women would expect to live in this way until they had given birth to one or more children. They would then be allocated their own field and began to cook in their own house, a transition which confirmed their position as wives. It also marked the moment at which they became responsible for managing their own food supply. If there were no other daughters-in-law, or if a young woman was slow to give birth, this period might last for five years or more. In other cases, it might be as short as a year. The *miaha* then gained the status of *dhako matin* ('junior woman'), which she had until menopause. At that point, she became a senior woman, *pim*. A *dhako matin* might still work on the older woman's fields and help her in the house, but this now depended on the quality of their relationship. The senior woman had no powers of compulsion. The norm in a good relationship seems to have been that the *dhako matin* would work on the senior woman's field several times a week. The senior woman, in return, would do some work on the *dhako matin*'s field. The *dhako matin* might still fetch water and firewood for the senior woman. Eventually, as her sons married, or her husband married more wives and she reached menopause, the *dhako matin* would herself gain the status of *pim* and get access to the labour of junior women. As one daughter-in-law gained her own field and cooking pot, another would marry-in. So, although a *pim* did not have sole access to the domestic labour of any one *miaha* for a long time and although she could not have more than one *nyar ot*, if she had several sons, she would have the labour of a succession of younger women at her disposal. They would also probably continue to work for her in her fields. In this way a woman could, in normal circumstances, expect increasing access to labour as her household subdivided.

Rising real wages in the 1950s and 1960s made it possible for a young married woman to live with her husband in town. Many women began to spend the first years of marriage with their husbands, rather than in their mother-in-law's house. Commonly, the woman would return when several of her children had reached school age, because it was cheaper to bring up children in the rural areas. By this time, she was *dhako matin*, with her own heavy load of farm and domestic responsibilities. She would not have much time to spare for her mother-in-law. The older women also had less authority. She and her husband would often be economically dependent on the younger woman's husband. Older women complained that daughters-in-law did not respect or care about the welfare of their seniors in the way that they used to. There did seem to be a difference between the amount of labour women with the status of *dhako matin* gave to senior women in the previous generation and the amount they themselves received from such women. Older women described the relationship in the following ways:

> 'I would work in my mother-in-law's field roughly every three days. I just went, without being called.'

> (Milcah Ambogo)

'I would help my mother-in-law in her field until the work was finished.'

(Rose Onyango)

'I would spend equal amounts of time in her field and my own.'

(Judith Mboga)

Younger women describe their work in their mothers-in-law's fields in quite different terms.

'I work in her field if she is sick.'

(Helen Ogendo)

'She may call me and, if I have finished, I help her.'

(Rispa Abiro)

Another reason for this difference was the widespread shortage of farm labour in rural households. Labour migration deprived older men and women of the labour of their sons before they also lost access to female labour. This loss was not so significant because the labour input of young men into farm work was not great and it was partly compensated for by a rearrangement of work tasks. Older men and young boys took over the herding responsibilities of young adult males. But the drain of labour power that followed the emergence of long-term migration increased the labour burden on adult women. A woman would be hard pressed to finish all the work needed for her own fields, quite apart from the time needed to make baskets or do some trading. She would usually have little time to spare to work for her husband's parents. For the same reasons, women from the same generation did not spend much time working on one another's fields.

To sum up: because of the decline of farming, off-farm income, particularly remittances, played a central role in household reproduction. Men dominated access to this resource because they were the ones who undertook wage labour, while few men were willing to allow their wives to undertake independent trading or cash cropping. These processes eroded women's spheres of authority and responsibility in the household. When a household's livelihood depended on men's ability to earn a wage, it tended to become more unified, under his authority. Ideological change both promoted and reflected these changes.

But the ideal pattern of male 'breadwinner' and dependent wife was far from the necessities of survival in most households.

Phoebe Osaro married in 1964, when she was sixteen. Her husband was a tailor in Shinyanga, Tanzania and she lived there with him until 1967. Then they both came to build a house in the *dala*. She had lived in Koguta ever since. Her husband moved to Njoro, in the Rift Valley, in

1972. Later, he came back to Koguta, and then moved again, to Nairobi, in 1987. He used to visit about twice a year, but he didn't bring any money when he came. He would send about 50 shillings every few months.

'I started trading in vegetables in 1975. I thought of this because I was having problems. Sometimes the man works outside, but doesn't send money because he drinks. I didn't have money to buy soap and vegetables. There was a woman in the *jokakwaro* who traded in vegetables and she advised me to do the same. I started off with 5 shillings that I got from selling baskets. In 1979, I visited my sister. She and another woman were trading in maize. She said it was good. It was better than trading in vegetables. The supply was low and there was a lot of demand. I asked them how much you needed to start. They said "Any amount you have". The first time, I talked to a Kipsigis man and got him to give me some maize to sell for him and I kept the profit. Kipsigis men do this because they don't want to sit in the market. I did this for six months. I saved some of the ten or twenty shillings profit I made every time and I made some baskets to add money to my savings. Then I started trading for myself. I bought a sack of maize. Now I buy between five and thirteen at a time and I trade on all of the three market days. I make a profit of 50–70 shillings a day. I'm fed up with maize trading, because it makes my body ache.'

What mattered was whether people could earn money locally, through trading, or farm labouring, or whether they could get access to food or cash from other people. All this meant that economic roles were becoming more variable. In some cases this amounted to a transformation of gender relations and household structures that challenged received gender stereotypes.

Where the husband kept up active contact, or even was resident, he found it extremely difficult to play the role of breadwinner. If a migrant, he often could not send remittances, or, if he was resident, there were few opportunities to earn a regular income. Some women had altogether lost access to their husbands' incomes, taking on full financial responsibility for themselves and their children. Where the man was a migrant, this de-coupling of the rural and urban components of households left the woman as the effective head of the household. It was a very vulnerable position.

Katherine Okonya's husband was a night watchman. He never sent money home. She had eight children and a field of one acre. Katherine's house was in a poor state, with a leaking roof, and she had almost no furniture. Katherine had one and a half acres of land and she had half an acre under crops in 1988. She prepared the land by hand, because she could not afford to hire a plough team. But the land was on the *siany*. It flooded easily and she lost the whole crop after heavy rain. A small crop of millet survived the flooding, but cattle ate it. They strayed on to her land when she went to Ahero to visit her mother, who was sick. Katherine had so many children that she had never been able to save money to use as capital to start regular trading. Her income was immediately swallowed up by her purchases of food. Katherine survived by working on other people's farms. As well as selling baskets, she was also given money, food and household goods by her kin. Her two teenage daughters had to drop out of school because she could not pay fees for them.

Unless she was a widow, the authority of a woman in this position was only provisional. Her husband might turn up at any time and intervene in her running of the household. She lacked the ability to acquire or dispose of the household's assets as she saw fit. It was a frustrating position to be in – responsibility without power. Most women put up with these frustrations because the alternative looked even worse.

Eunice Obala was born around 1930 at Kamgan, on the Nyabondo Plateau. She went to a mission school in Kakamega. She didn't want to get married. She wanted to be a nurse, but her father refused. 'I was his beloved daughter and he had named me after his friends. I had no choice but to go along with what he wanted, but it was very painful for me.' She married a JaKoguta, a JaRamogi, in 1946. 'The bridewealth was brought, so I agreed to the marriage. I would have liked to marry a different man, but finding one would have involved walking around and I didn't want to do that.'

She lived with her husband in Nairobi. 'One day, I got a letter telling me that my father was ill. I wanted to go home, but my husband refused to give me money, saying that there wasn't any. Then I got a letter saying that my father had died. I wanted to come, but my husband refused again. Things got very hot.

'A neighbour came and found me crying bitterly. He gave me money "to buy soap" and I used it to come home. He gave me ten shillings. It cost eight shillings to get to Kisumu and two shillings to come here. I came home only with my body. My husband took all my things. I found that my beloved father was already buried. That is why I left my husband.

'I was at home for five years. I worked at Nyabondo hospital, cooking for the patients. All this time, my brothers said that I was *migogo* (a returned wife), so where would I be buried? Many men came looking for me, but I didn't want to marry, because of what had happened the first time. I wanted to do whatever I liked, with nobody stopping me.

'My brothers negotiated with a man from Kano, who brought some cattle. But even there I lived with trouble. I lived there for nine years and then my husband turned nasty. He married several more women and he sent me away. He told me that he didn't want me at all. I had two children with him. One died. I took the other one with me. One child had also survived from my first marriage, but I could not take her, because of Luo custom, because bridewealth had been paid.

'I came home. This time my brothers kept quiet, because I had been chased away and they knew I was telling the truth. My mother was sick and I helped her very much. I worked as a Maendeleyo ya Wanawake teacher, teaching reading, writing and basket making.[7] When I was in Nairobi, I got to know a European woman who taught me spinning and weaving. I did not want to forget it, so wherever I was, I would start this. After six years, I came to an understanding with the old man who lived here. He was a widower. I wasn't happy about marrying him – I was counting what had happened. I just wanted somewhere to be buried.' He died in 1978.

Eunice's daughter was living with her. Eunice had a farm of one and a half acres. 'My husband's son took the title deeds. I do not know his character.' She no longer did any teaching, because her health was poor. 'I have no clear way of getting money. I have to farm and I can't do two things at once. I leave the weaving to my daughter.'

The rate of divorce in Luoland has usually been thought of as low compared with some other ethnic groups in Kenya (Potash 1978). The main reason was believed to be the consequences of divorce for Luo women. These reflected the strongly patrilineal nature of rules governing descent, marriage and the inheritance of property. A woman married into the patrilineage. Bridewealth secured rights to her labour and over the children of the marriage. Women who left their husbands would expect to lose their children. They might also not get a warm welcome in their natal homes. A divorced woman was a burden to her family, both because they might be required to repay her bridewealth and because they would need to provide her with land. Nevertheless, marriages did break down. I was told that many women had left their husbands and 'gone to town' because they were not getting any support. There were a few such women in Koguta, as well as young women with children who had not married. In Kombewa, in the north of Kisumu District, Odaga found women who had earned the epithet '*Odhi, oduogo*' ('she leaves and comes back again'), as they oscillated between an unsupportive husband and their already overstretched parents (Odaga 1990).

If a returned migrant did take his responsibilities seriously, his options were very limited. Many older men had hoped to get support from their children when they retired, but not many did. Other returnees were younger men who had lost their jobs, or never found work, in town. The pressing need to find money led some men to try to intensify farm production and sell some of their crops. The effects of this strategy on gender relations resembled the changes that had been played out in better-off households in the 1930s and 1940s.

'I want to pull her up to catch what I'm thinking.'

> Aggrey Obala was in his late twenties. He had some training in agricul-
> ture and had worked on a big farm in Kiambu. Aggrey decided to come
> back to Koguta and start farming when he was in his early twenties. His
> father gave him three small fields, half an acre in all. He terraced the land
> and dug it over to remove the stones. He planted maize, eleusine, fruit
> trees, vegetables, potatoes, cassava and sweet potatoes. He also had two
> beehives. The farm was his only source of income and he put a lot of
> effort into it. What he wanted to do was buy more land. Very few
> JoKoguta would be willing to sell, so he would move to South Nyanza
> and buy land there. Aggrey stressed that he and his wife, Mary, worked
> together on the farm. They made decisions together about planting the
> crops and they both did most of the jobs on the farm. Their main cash
> crop was tomatoes. Mary sold them: 'She brings the money and I tell her
> to keep some.' They both had rights over this money 'En gi teko kendo
> an gi teko.'[8] Aggrey took this attitude because he wanted 'balance' in the
> marriage. 'I want to pull her up to catch what I'm thinking. She is
> coming.'

Aggrey was serious about including Mary in the management of the farm, but their relationship looked and sounded far from equal. However, Aggrey's father still criticised him for giving Mary too much say in their affairs: Aggrey saw himself as breaking away from a norm of male control and female subordina-tion. That his limited attempts to do so attracted criticism shows how much authority men who farmed expected to have. The contrast with the authority women had expected to have over their own fields in the past could not have been greater.

The idea that it was right for the man who took his family's welfare seriously to take charge reflected the ideological legacy of labour migration, as well as the shrinking resource bases of rural households. A man living at home who tried to feed and clothe the household was a good husband and his involvement in farming was seen as a help to his wife, not an unjustified takeover of her sphere. Men who were involved in farm work described themselves, and were described by women, as 'helping' (*konyo*). *Konyo* gave entitlement to a role in farm deci-sion making and a share of crop income. It was then only a short step to full appropriation, which could be justified by the man's greater financial responsibilities and his ultimate ownership of the land. An older woman had more authority to resist full appropriation and work out a compromise. Her husband and she might share the crop income and the financial responsibilities. Younger women who were still seen as newcomers to the community, could not easily adopt that position. Yet even a joint strategy was often not sustainable.

If a man was not prepared to pull his weight, either by running the farm or by trying to find a steady income, the woman had to do so instead. When this happened, she took on more of the decision making and the financial responsibilities.

When a man fell short of expectations about a 'good husband' his wife took on most day-to-day financial responsibilities by default. The choices open to a woman who took on these responsibilities were very narrow and the threat of a fall into extreme poverty was always in the background. It was difficult to build up capital for large-scale trading or sinking money into purchased inputs for the farm, because any profits needed to be used straight away to buy food. Even if she did manage to earn money, a woman became more vulnerable to extreme poverty as she got older and her health deteriorated.

The fact that most households could not hope to satisfy their need for food from farming threw up a contradiction. Many men hoped to provide their wives with enough money to allow them to remain at home, but this hope was becoming unrealistic for most people. Women needed to find money themselves.

Hannah Akech got married in 1971, when she was nineteen. Her husband worked for Kenya Railways. When they married he had a job loading and unloading trains. He got a job as a clerk and then as a station master.

'When I began to trade, my husband said it was a waste of money. The high price of maize and the transport costs would mean that the profits would be very small. He also said that I would run off. On the first trip I came back with KShs.600. He told me to give him the money. He said the money would make me leave him. He didn't do anything with this money for a long time. Then he used it to buy four sacks of maize for me to sell. I made KShs.1100 and he took the money again. Another woman who was trading advised me to give him the money. Then he would see that I wasn't wasting it or going astray with it.'

Hannah moved to Koguta in 1988 so that the children could go to school there. She wanted to start trading in maize again, but she would not tell her husband straight away. 'I'll start by using some of the money he sends me. If he knew I was trading, he wouldn't bring me any money.'

(Hannah Akech)

Hannah's husband's ambivalence was common. Although many men disliked the idea of their wives having their own money, they often changed their minds. Many women who traded commented that their husbands had not at first liked the idea, but when they saw that they could bring home money they no longer objected. Women who traded regularly often had much more authority. Their command of a cash income gave them a greater voice (*duol*). A *dhako ratego* played a much greater role in making strategic decisions (about buying and selling assets, about what crops to plant, about the children's schooling). Her voice carried most weight when her money financed these activities. This greater authority seemed to come both from sheer financial clout, and from the assertiveness that came along with an independent income.

> 'What the strong women have in common is that each has been in a position where she several times asked her husband for money to buy food and he replied that he didn't have any; that she should start making baskets. There aren't any women with husbands who don't help who have done nothing to get money. You will work hard only if you don't have money. Women who are like this are stronger than other women, but they can't be equal to their husbands – the husband should still be over his wife.'
>
> (Martha Odhiambo)

Whether or not a woman could hope to make her own living depended very much on her age. It was difficult for younger women to spend most of their time trading or doing paid farm labour, because of their responsibilities for childcare. Younger women also had less land, because their husbands had inherited a subdivision of an already small piece of land. Older women might be freer to trade or do farm labouring, but their need for money was often greater, especially if they were paying school fees. Trading was a tiring and frustrating way to make money, as Judith Mboga's story showed (see pp. 149–50). Only a few women became regular traders. Far more put a livelihood together by juggling several different activities.

Odaga gives an insightful analysis of similar processes in Kombewa Sub-location, in the north of Kisumu District. Like Koguta, Kombewa is an impoverished area where low agricultural yields and land shortage make access to off-farm income critically important for survival. Odaga had some unexpected findings that revealed recent processes of agrarian change in the region. In an area of acute land shortage, she found that 47 per cent of the women in her sample did not cultivate their own plots in the short rains of 1986–87, or reduced their labour into them.[9] She also found that 42 per cent of her sample were not receiving remittances of any kind. Taken together, these figures offer clues to the strains facing women and men and the strategies open to them. Women invested their labour where they were likely to receive most return. The household head did not automatically control women's labour. In households where the husband was living at home, there were circumstances where this calculation would encourage women to co-operate with their husbands in a 'household strategy'. This might involve the woman working with her husband on the farm to produce crops for sale (over which he had decision-making power), or he might work the land while she engaged in paid agricultural labour or trading in order to bring in a cash income. This was a risky strategy for the woman, because she could not be sure that the crop income would be used for the benefit of the whole household (cf. Heald 1991, discussed in Chapter 4). Where it failed, the woman might withdraw her co-operation and follow an individualised strategy of labouring or trading, rather than working on the household plot. Women's ability to engage in income-earning activities depended on the access of other members of the *ot* or *dala* to her labour, but this, in turn, was contingent on their providing her with some security. Because this part of the bargain was becoming increasingly difficult for men to fulfil, the material basis of household co-operation was being undermined. As in Koguta, interdependence was only one approach to constructing household livelihoods. Fragmentation was becoming more common.

Conclusion

Relationships within households in Koguta, and the ideologies that legitimated them, were changing under the strains of impoverishment. Growing numbers of men, migrants and returned migrants, could not hope to meet their households' needs for money. Decades of labour migration had created the expectation that men and women follow a division of responsibilities in which the husband acted as the 'breadwinner', but economic necessity made this division contentious and often unsustainable. New divisions of responsibility, sometimes amounting to new household forms, were appearing. Households with no male head, resident or migrant, were becoming more common. There were households with an older woman head, with unmarried or divorced daughters, and their children, and households where an older woman was fostering the children of an unmarried or divorced daughter based in town. But male involvement in finance and decision making covered a spectrum from considerable control to desertion and no clean break could be made between male-headed and female-headed households.

In households where husbands' remittances and visits were so irregular that the wives were effectively household heads, male involvement was often intermittent; women's authority was only provisional. This applied even more when the husband was living at home but not contributing much to the household's livelihood. Negotiating spheres of authority was a delicate business. Women gained access to most productive resources through their relationships with men, whether husbands or sons, and their power over these resources was temporary and subject to challenge. Another trend was towards appropriation of women's labour-power by men who were trying to make a living from farming.

The one avenue to accumulation open to women acting on their own account, trading, was limited by low demand in the local economy; by women's farm and domestic workload; and by men's hostility to what they saw as a loss of control over women and a disruption of the rural household. Many rural women saw their economic responsibilities increase but found it almost impossibly difficult to fulfil them. The less a man could contribute to the household's livelihood, however, the weaker was his case against his wife's trading. Some men acquiesced or even encouraged their wives to start trading. If the woman became a successful trader, her domestic power would be likely to grow with her responsibilities. She might challenge the constraints that so restricted other women. Women like this were a small minority and trading was such a low-earning and risky activity that few could hope to emulate them, though their example might legitimate more widely the growing need for women to look for off-farm sources of income.

The gap between expectations of men and their ability to deliver led to tension and conflict.[10] It also raised questions about domestic responsibilities and authority. People responded to these questions in quite different ways. If the husband was a migrant, or if he was living at home without a regular income, the weakening of financial links between husband and wife raised difficult issues concerning the wife's response and her husband's right to control how she responded. Men were often reluctant for their wives to begin trading, seeing women's greater financial independence as a threat to their authority and to the unity of the household. The conflict could go in several different directions – the husband might acquiesce; or he might himself appropriate his wife's labour in order to grow crops for sale. These outcomes were sustainable only through acknowledging interdependence and a willingness to negotiate. Whether they were sustained could not be predicted from material conditions alone. The outcome involved questions of personality and agency. Where neither of these outcomes could be reached, the husband might disengage, leaving his wife responsible for supporting the household. Even then her authority was provisional and subject to challenge, either from him, or, more unusually, from close male kin in the patrilineage. Changes in the regional political economy, particularly the decline of the remittance economy, were making gender relations based on simple dependence unsustainable and were increasing pressures for negotiation and bargaining. Where this failed, the centrifugal pressures on the household as an economic unit were immense. What held it together was women's need for marriage to get access to land and to retain custody of their children. The South African cases in Chapter 4 showed how the disappearance of the material basis of households through land shortage and unemployment can pull marital relationships apart, or deter people from marrying in the first place. The migrant labour economies of Kenya may grow more similar to these regions, as agricultural land becomes scarcer and job prospects shrink in the urban economy.

Notes

1   For a discussion of the role of changing ideas concerning the roles of men and women in Kipsigis society, see von Bülow 1992.
2   *Puodho* – field; *Lowo* – land.

3   Cf. Margaret's comment on p.156, 'Our friends come to visit us in cars.'
4   The Odhiambos took part in only the first two of the three rounds of the budget survey, so they are not included in Table 5.4.
5   Literally – *bayo*, to wander or walk around, also has sexual connotations.
6   In fact informants used the term *wendo* more often.
7   Kenyan national women's development organisation.
8   'She has the power to decide and so do I.'
9   Rainfall in the region is bi-modal, with two growing seasons in the short rains (October to November) and long rains (March/April to May/June). Odaga found that cultivation was more intensive during the long rains.
10 Silberschmidt (1992) found similar tensions in neighbouring Kisii District.

Conclusion

'I'm here because when you have a family, you should try different places. When you have a family, they shouldn't be clustered in one place, because when they die, they all die. When they are in different environments, trying to make a living, they won't all die at the same time.'

(Interview with Ladlong Cornelius Masire, Geysdorp, North West Province, South Africa, 1 June 1999)

At the beginning of the book we saw how severe the human impact of Africa's economic problems has been. Data on incomes and poverty make grim reading, and look even worse when set against the economic growth seen in many other developing countries since the 1960s. The rest of the book is concerned with the local histories and livelihoods that lie behind these national-level figures. Chapter 1 outlines the ways in which rural households in Kenya and Tanzania have been drawn into intensive commercial farming. In Chapter 3, we saw how in Northern Zambia, Southern Zimbabwe and Lesotho, agriculture has become at best a backstop for households whose main source of money has long been the wages of their members who found work in the mines, farms and cities. The story of economic decline in Koguta shows how this came about. Several generations of rural households made a judgement that their best hope lay in finding work in the urban areas and in educating their children to improve their chances in the labour market.

Even this picture of local diversities involves painting with a broad brush, however. There are great diversities within local economies. Chapter 1 showed how severe poverty can co-exist with intensive commercial farming in a region like Lushoto in Tanzania. There, the very poorest people, mostly women, were those denied the chance to look for work on the farms of the better-off. There are thousands of landless households in Central Province, Kenya, while many people farming there do not grow the high-value crops. Some people living in migrant-labour areas try to farm intensively. In Koguta, some households managed to grow coffee, tomatoes and onions for sale. Nor was everyone equally poor. There is a big difference between people who can eat meat three times a week and people who would be lucky to eat it once a month.

Explaining these diverse histories requires an understanding of how large-scale social and economic changes were played out at the local level. These include the ways in which localities were incorporated into colonial economies, the effects of post-colonial state policies, the operation of markets and the activities of international capital. Analysing these processes involves exploring how they have meshed with local strategies for making a living. Some strategies are oriented towards accumulation and upward mobility. These strategies are shaped by the opportunities people find around them, but they also reflect their values and cultural orientations. Some farmers in Lushoto followed pre-capitalist accumulation strategies, acquiring labour for their farms through polygyny and having many children. Others followed a quite different strategy, based on employment, investment in higher-value crops and land purchase. They put a high value on education and invested in off-farm enterprises. The *Josomo* from Koguta made a judgement early on in the colonial era that the road to upward mobility lay through education, rather than through farming.

Local strategies for making a living also include all the ways in which people have tried to stave off impoverishment. Mambwe villagers in Zambia used the border with Tanzania as a resource, taking advantage of cross-border differences in demand for commodities. They also activated kinship links across the border to facilitate trading. Bemba people drew on their fluid kinship relations to keep their options open as they moved from dependence on labour migrancy towards more locally based livelihoods.

The importance of understanding what drives local strategies is underlined by the history of states' attempts to create commercial farming from above. In post-colonial Kenya, many large farms which had been transferred intact were underutilised, their owners' attention fixed on investments in the urban economy. The medium-sized farming sector created by land reform suffered similar problems. This story was repeated in former Bophuthatswana in the 1980s and 1990s, as many beneficiaries of Mangope's land redistribution neglected their farms. When rural households can combine farming with off-farm activities, and can get access to markets, smallholder commercial farming can emerge from below. The links between farming and off-farm activities become a source of dynamism, rather than putting a brake on economic growth. Governments intent on creating full-time farmers through land reform programmes need to take seriously the fact that when African rural households have a choice, they prefer to spread risks by following multiple livelihoods.

While one theme running through the book is diversities in rural livelihoods, another is the commonalities between localities in the predicaments people face. We have seen how widespread has been the downturn in urban economies and how this downturn has reduced the demand for the migrant labour which rural households have depended on for their livelihoods. Another way of looking at this problem is to see it as a massive increase in the risks rural households face as they attempt to construct livelihoods. There is nothing new about the prevalence of risk in rural Africa. Most of the continent is subject to unusually great climatic variation. In parts of Ethiopia, Tanzania and Zimbabwe, there is roughly 10 per cent probability of near total crop failure (Collier and Gunning

1999; Kinsey *et al.* 1998). African farmers have long been subject to volatile commodity prices. Depending on migrant labour has always been a risky strategy. Most migrants at most times did low-paid, insecure work. However, not being able to depend on a wage income leaves rural households to make a living from even more insecure and low-income activities. Risk and low returns drive people to diversify their livelihoods, as is happening across the continent. We have seen that multiple livelihoods are far from being a new strategy. What does seem to be new is the extent of diversification. It is a well-established coping mechanism, but it leaves rural households, at best, treading water. Diversification takes energies and resources away from any particular activity people may be engaged in and reduces people's willingness to innovate, other than by diversifying into new sources of livelihood. The aggregate impact of diversification, therefore, is to reduce still further the growth potential of African economies.

Like Ladlong Masire in Geysdorp, households may disperse across several different environments, to avoid the risk that all their sources of livelihood fail at the same time; they may try to combine wage work with farming or they may do some small-scale trading. These various livelihood strategies throw up different issues for members of a household. Commercial farming rests on inter-dependencies. Men need access to women's labour. Women want access to the income the crop brings in. These interdependencies do not in themselves guarantee an equitable outcome. They may encourage men to take over decision making, as had happened among Donna Pankhurst's informants in Murasi, Zimbabwe. They may lead to stalemate, with women withdrawing their labour from the commercial crop, as happened to tea growers in Kericho, in Kenya. Or, as among the Teso tobacco growers, women who believed that they would benefit from the crop income readily co-operated with their husbands. All this looks more open than models which assume that African households are less corporate than those found in South Asia. Bargaining, dictatorship, conflict and co-operation are all possibilities. None are inevitable. Nor are they a fixed architecture for gender relations. Successive generations of JoKoguta have seen households become more corporate as they became dependent on men's ability to earn a wage and then less corporate as the burden of household reproduction has been thrown back on to women.

Diversified livelihoods can reduce the interdependence of household members. The tensions this leads to can make it more likely that households fragment. Poverty, and diversified livelihoods, by undermining interdependencies, may make it less likely that men and women combine to form households in the first place. There is some evidence from South Africa and Lesotho that this is happening, as men find it more and more difficult to provide the money needed to support a household and women decide that they may be better off taking on this responsibility themselves. In Koguta, as in many other of the more densely populated parts of Africa, the next generation will be virtually landless, their options little different from rural dwellers in South Africa. In regions like this, the material bases of household formation, household structures, divisions of responsibilities and norms of behaviour are all under strain as people contend with the many different processes pushing them deeper into poverty. Where land is less scarce, many households lack the means to work it intensively because it is increasingly difficult to get access to capital for farm investment through the labour market. Risky environments also reduce their inclination to concentrate their efforts on intensifying production. Governments need to recognise that diversification is a rational response to risk. It is their responsibility to create more predictable environments and reduce the risks that drive people towards diversification. Until that happens, people in rural areas will have to maintain the immense ingenuity that making a living in rural Africa demands.

Appendix A
Research methods used in the Koguta case study

I carried out research in Koguta from November 1987 to February 1988 and from May 1988 to March 1989. Throughout both periods, I lived as a member of a Koguta household. I became reasonably competent in Dholuo, though I always worked with a research assistant, Ruth Waga, whose help I often needed for more complex discussions.

My hosts were members of a lineage which had early on been involved in the mission churches. Members of this lineage had been among the first people in the area to go to school. A large number of these people had sent their children to secondary school and several had gone on to higher education and professional jobs in the urban areas. In order to study the widest possible spread of household types, I extended my study to three other lineages whose members did not have a history of early involvement in Christianity and education.

I used several research methods, chiefly a survey of 104 households (using a random sample), in which I collected basic socio-economic data, 60 life histories and a budget survey of 24 individuals in 19 households; and I carried out interviews which were focused on specific themes (such as the sexual division of labour in agriculture). I also had many informal discussions.

Budget survey

The aim of the budget survey was to provide quantitative data about the economic behaviour of different types of individuals. Economically active household members were questioned separately in order that budgetary roles and transfers of resources within households could more easily be studied. In order that non-market material relationships between households could be illuminated, informants were also asked about transfers given and received during the survey period. This information is notoriously difficult to quantify, but it probably does give an indication of orders of magnitude.

Twenty-four individuals in nineteen households were questioned retrospectively about their incomes and expenditures, in cash, and, as far as they could recall, in kind, in the previous month at three points in the year 1988/9. In order that they reflect seasonalities in incomes and expenditures, these points covered the hungry season just before the long rains harvest (May/June); the month immediately after the long rains harvest (August) and the month when the largest tranche of school fees falls due (January). Informants were also asked about large receipts and outgoings of cash in the intermediate periods.

Life histories

I originally intended to collect around twenty life histories as a supplement to a larger number of interviews on specific topics. I finally collected sixty usable life histories, because I found this kind of interview to be a flexible research tool which also encouraged many informants to open up far more than in other types of interview. Sixty life histories gave a good spread of informants according to the criteria of age, sex, household structure, economic activities and socio-economic position.

Sampling

The samples were chosen in order to cover individuals in different types of household, according to the criteria younger/older, migrant male head/non-migrant male head/female head, better-off/poorer, in order to observe the widest possible variety of behaviour. Because one of the aims of the study was to synthesise an historical and contemporary analysis of economic and social change in Koguta, there is substantial overlap between the informants whose life histories were collected and those who took part in the budget survey.

There is no a priori link between qualitative research and non-statistical sampling techniques, but the subject of my research required that I talk to individuals in a wide variety of socio-economic positions in order to explore a range of social processes and behaviour. Some types of individual might not have been picked up in a statistically derived sample. For example, there was only one university educated returned migrant in the village where I lived, but interviewing him was important for understanding processes of social differentiation. Equally important was the fact that the unit of analysis changed from topic to topic in the study, because I was looking at processes going on at the levels of the individual, the household, the compound and the lineage, as well as long-term processes of class formation across generations. In many cases I did not know what the appropriate unit of analysis was at first, for example, in studying the material aspects of kinship links, this is precisely what I was trying to discover. The process of selecting informants was thus largely additive, as further criteria of selection became relevant, and as I met people who were interesting for my study (and also sometimes subtractive, as some people who initially seemed potentially good informants proved not to be particularly enlightening).

Themes covered in the life histories

One of my main research interests was in long-term analysis of class formation and changing gender relations during and since the colonial period. In order to examine these processes, I needed to collect information about individuals' economic activities and relationships, particularly with regard to access to and control over key resources such as land, labour, education and money. Topics for research included patterns of accumulation and impoverishment, relationships within the household, the compound and beyond and changing rural-urban linkages.

Because, in many cases, I was dealing with a long time period (sometimes over seventy years), I did not expect to be able to collect accurate numerical data, particularly for the early decades. I considered that information about behaviour and social processes was less likely to be inaccurate, although it does present its own difficulties.

Most interviews lasted for around two hours, although I discussed their life histories with some individuals in more than one interview. I collected some basic information from all informants, namely, age, or approximate age; father's occupation and approximate size of his landholding and herd; education level; marital, migratory and occupational history of self and spouse; current landholding and land bought or sold; stockholding; crops currently grown and sold and long-term changes in these; investments; current economic activities and sources of income; children's education and occupations; and whether the children provided economic support.

I structured the interviews around likely major life events. For a woman, these included marriage and the consequent move to her husband's home; getting her own farm and cooking pot; and the final return of a migrant husband. For men, these were the first migration and subsequent jobs; marriage; and retirement. There were then several recurring themes.

Some were relevant for certain informants, and others were not. I asked a number of women about their relationships with senior and junior women in their compounds over the course of their lives, while I asked many men and women about the husband's involvement in farm decision-making while he was a migrant; I attempted to establish how prosperous male and female traders had achieved their position and why some elderly widows wielded economic power in their compounds, while others did not. Many topics emerged only during the interviews and when an informant made a remark which set off a new line of enquiry that seemed important, I followed it up. Some individuals were more informative, or readier to talk about some topics than others, so it was often a question of getting a feel for how the interview was going and acting accordingly.

Problems

Many older informants were vague about dates and ages, since these were not important pieces of information until recently. I used various methods for establishing ages and dates. People often knew whether they were a little older or younger than other individuals who appeared to be of the same generation and who, because of greater exposure to education, knew their dates of birth; they could also tell me whether they were born, married and so on near a well-known event of which I knew the date (for example, the coming of the first missionaries; the World Wars; a major famine in 1931–32; Independence), or whether they were old enough to look after cattle at that time (probably over seven years old) or had given birth to their first child (probably around seventeen years old), and so on. In this way, I could usually place an event to within around five years.

The other problems were harder to avoid. Informants sometimes telescoped the past, giving a description of an undifferentiated 'long ago', in which it was hard to distinguish the 1920s from the 1940s, or they quoted apparently exact figures from one period to cover a much longer one (for example, giving their last earnings from employment as the typical wage in their career). I tried to get round this difficulty by asking informants about specific periods ('when you were first married'; 'at the time that your husband was working in Mombasa', and so on), but this was only sometimes effective. It was easier to derive a periodisation from some informants than others and for certain topics than others. Men could usually remember the dates of entering and leaving various jobs, so it was not difficult to periodise a migrant's career, but gradual changes in labour use in a rural household were harder to track. One is forced to fall back on, for example, people's memories that when they were young, most women worked on their gardens in the morning and afternoon, whereas now they farm in the morning and make baskets for sale in the afternoon. Prior knowledge of the changing political economy of the region (growing cash needs; land shortage; distance from markets for cash crops) was necessary to interpret such observations.

Another source of difficulty was the possibility that informants' views of the past were coloured by nostalgia, or that they were projecting contemporary patterns of behaviour back into the past. These pitfalls are intrinsic in the collection of oral history, but they can be at least partly avoided, as in the comparison of the relationships between senior and junior women in successive generations. The aim was to find out whether, and why, the access of senior women to the labour of junior women had diminished. Most older women said that they had helped their own mothers-in-law more than their own daughters-in-law now helped them, but this perception could easily have been mere nostalgia. Further evidence was needed that there had, indeed, been a change and this came from the terms in which older and younger women described their relationships with senior women. Asking the informant a more specific, and less emotionally charged question than a direct comparison between generations, produced answers that I could then myself use to make comparisons.

Perhaps the most difficult problem to deal with, however, was the possibility that an informant was withholding information, or not telling the truth. There were some obvious strategies for lessening this possibility. I always began interviews with innocuous topics and slowly worked my way towards more sensitive subjects (this is also how Luo people behave when they want to discuss a delicate topic) and I tried to introduce potentially embarrassing questions crabwise. For example, many people are reluctant to admit to having little, or no education, so I often prefaced a question about educational attainment with the question 'Were there many schools in your community when you were a child?' The informant could then explain that there were very few schools and the admission that he or she had received little or no education was easier to make. Outright lying could sometimes be spotted because the information was inconsistent, or implausible. Because this was such a small community, I was often able to cross-check information through discreet enquiries.

I have been able to use the material collected in the life histories to analyse long-term processes of economic differentiation, because the principal source of economic mobility within and between generations has been the labour market. Qualitative distinctions can often be made between households with regard to the labour market position of their members (unskilled, artisan, clerical, managerial employment, and so on), while differences in labour market position have generated quite distinct trajectories of household development. For example, it was possible to see how some individuals were able to use resources gained through trading or better-paid wage employment to invest in their children's future labour market position through funding their education to secondary-school level and beyond, and how others were hampered from doing this by low wages. It would have been more difficult to collect material on very long-term changes in economic position in a community where mobility involved primarily quantitative changes (land holding, cattle holding and so on), which informants are perhaps more likely to misremember.

Asking other people to tell one the story of their lives is a highly artificial undertaking. A 'life story' is an intellectual construct whose structure and content reflect the priorities of the researcher and the images the informant projects back into the past as much as tangible realities. Despite this artificiality, I consider that it is possible to collect reasonably accurate material about certain topics in an interview that is structured around the chronology of an informant's life, provided that it is not done naively.

Appendix B
Luo segmentary lineages

Anthropologists who have studied Luo lineage segmentation give conflicting accounts of its structure (Evans-Pritchard 1949; Southall 1952; Wilson 1961; Whisson 1964; Pala 1977; Parkin 1978; Shipton 1979, 1985, 1988, 1989; see also DuPré 1968). The discrepancies stem from inconsistent use of English terms like 'clan' and 'lineage' and from the fact that segmentary groups and segmentation have been more fluid and geographically variable than was first believed. Some generic terms also vary with the context. In his study of land rights in Luoland, Shipton argues that Wilson's account of Luo lineages and land rights is the most thorough. The largest subdivisions[1] Wilson distinguishes were between *ogendini* (sing. *oganda*) and there were about thirteen of these in what became Central Nyanza District at the time that the British arrived.[2] When the colonial administration was established, most were used as the basis for demarcating locations. In most *ogendini* there was a chief, *ruoth*, whose power was variable, but who often had powers of mediation. In some *ogendini*, chiefship was routinised, with inheritance passing between close male agnates; in others, the situation was more fluid. *Ogendini* were subdivided into *dhoudi* (sing. *dhoot*). *Dhoudi* were normally exogamous (as were many *ogendini*), with a fairly clearly defined territory, *gweng*. Shipton argues that they were the highest level of grouping in which lineage largely coincided with territory (Shipton 1979: 158).[3] A *dhoot* was divided into *libembini* (sing. *libamba*) and these were the largest descent groupings within which individuals could freely exchange plots.[4] They were also the largest groupings whose members would sacrifice together frequently (Shipton 1979). Whisson suggests that some segments of a certain size or degree of segmentation would also co-operate in emergencies, while shedding blood would be considered ritually dangerous. The next level of segmentation in Shipton's account is the *keyo* (pl. *keshe*). He points out that the term also refers to an extendible straight strip of land occupied by a descent group and that it was at this level of medium-sized groups that settlement patterns most closely followed the principle of land occupation in strips. Although elders acted as representatives of their grouping at each level of segmentation for dispute settlement, Wilson suggests that it was the smallest unit to have an organised council of elders to settle internal disputes. Whisson considers that groups at this level would expect to co-operate in garden work.

These higher levels of segmentation are still important. They underlie local politics and struggles over access to state resources. Together with the links constructed by marriage, they shape the social networks that stretch across Luoland and the urban areas of Kenya.

Below this level were (and still are) small agnatic lineages, *jokakwaro*, which vary in genealogical depth, but they usually denote the next largest grouping above the single compound (*dala*, pl. *mier*, sometimes *pacho*). The *jokakwaro* consists of closely related *mier* who tend to be settled on contiguous strips of land that are subdivisions of *jokakwaro* land. This in turn is a subdivision of *keyo* land. In pre-colonial Luo communities, *mier* in a *jokakwaro* often herded their cattle together and they were joined in relationships of reciprocity in food and garden labour and by networks of bridewealth donation and receipt. A young man might receive cattle towards his bridewealth from other men in the *jokakwaro* and an older man might allocate some of the bridewealth cattle received from a daughter's marriage to other men in the *jokakwaro* who had donated cattle to him in the past. Informants related that the *jokakwaro* was the unit within which reciprocal relationships of this type were most common and social interaction most intense.

There are two important points to bear in mind here. This account of lineage and terms and structures gives a rough picture of kin groups. However, it should not be read as a rigid structure. People's use of kinship terms depends on context and the connotations these different words carry. The same person might be referred to as *JaKoguta*, or *jalibamba* might also be correct, but *JaKoguta* stresses one's common membership of the kin group with the other person; *jalibamba* stresses that the person belongs to another, possibly rival sub-clan.[5] People also used particular kinship terms as explanations for their relationships with others. The term *anyuola* has this kind of flexibility. It means a group of closely related kin. However, while saying, '*Wan joanyuola achiel*' ('We belong to the same *anyuola*') stresses common membership of a close kin group, it can also be used to make a statement about identity and moral obligation. In this sense, somebody might say this to stress a bond with any other *JaKoguta*. It is also vital not to overlook the importance of relationships traced through women, even though the 'official' lineage structure makes these invisible. Affines may help you find a job or a place to stay in Nairobi. Because clans are exogamous at the local level, affines also broaden your networks out all over Luoland.

Notes

1 Apart from a broad division between people of West and East Luoland (Parkin 1978).
2 Central Nyanza was later divided into Kisumu and Siaya Districts.
3 Higher-level units were less often activated.
4 Whisson (1964) states that these were also referred to as *anyuola*.
5 Odiyo (p.134) used the term *jalibamba* to describe the man whom he ousted from his land in a court case.

Bibliography

Adams, J. (1991) 'Female Wage Labour in Rural Zimbabwe', *World Development* (19): 2–3.

African National Congress (1994) *The Reconstruction and Development Programme*, Johannesburg: Umanyano Publications.

Allen, T. (1998) 'From "Informal Sectors" to "Real Economies": Changing Conceptions of Africa's Hidden Livelihoods', *Contemporary Politics* 4(4): 357–73.

Amin, N. and Bernstein, H. (1996) 'The Role of Agricultural Co-operatives in Agricultural and Rural Development', Johannesburg: Land and Agriculture Policy Centre (typescript).

Andersson, J.A. (1996) 'Potato Cultivation in the Uporoto Mountains, Tanzania – An Analysis of the Social Nature of Agro-technological Change', *African Affairs* 95(378): 85–106.

Anker, R. and Knowles, J. (1981) 'An Analysis of Income Transfers in a Developing Country: The Case of Kenya', *Journal of Development Economics* 8: 205–26.

Arrighi, G. (1970) 'Labor Supplies in Historical Perspective: A Study of the Proletarianization of the African Peasantry in Rhodesia', *Journal of Development Studies* 6(3): 197–234.

Ashforth, A. (1990) *The Politics of Official Discourse in Twentieth-Century South Africa*, Oxford: Oxford University Press.

Baker, J. and Pedersen, P. (eds) (1997) *Rural-Urban Dynamics in Francophone Africa*, Uppsala: Nordiska Afrikainstitutet.

Bank, L. (1994) 'Angry Men and Working Women', *African Studies* 53(1): 89–113.

Barnes, T. (1992) *To Live a Better Life: An Oral History of Women in the City of Harare, 1930–70*, Harare: Baobab Books.

Bates, R. (1981) *Markets and States in Tropical Africa*, Berkeley and Los Angeles, CA: University of California Press.

—— (1989) *Beyond the Miracle of the Market: The Political Economy of Agrarian Development in Kenya*, Cambridge: Cambridge University Press.

Bayart, J.-F., Ellis, S. and Hibou, B. (1999) *The Criminalization of the State in Africa*, Oxford: James Currey and Bloomington, IN: Indiana University Press, for the International African Institute.

Beall, J. and Kanji, N. (1998) 'Households, Livelihoods and Urban Poverty', background paper for ESCOR-commissioned research on Urban Development: Urban Governance, Partnership and Poverty, draft manuscript.

Becker, C., Hamer, A. and Morrison, A. (1994) *Beyond Urban Bias: African Urbanisation in an Era of Structural Adjustment*, London: James Currey.

Becker, G. (1981) *A Treatise on the Family*, Cambridge, MA: Harvard University Press.

Beinart, W. (1984) 'Soil Erosion, Conservationism and Ideas about Development', *Journal of Southern African Studies* 11(2): 52–83.

—— (1989) 'Introduction: the Politics of Colonial Conservation', *Journal of Southern African Studies* 15(2): 143–62.

Berman, B. (1990) *Control and Crisis in Colonial Kenya: the Dialectic of Domination*, London: James Currey.

Berman, B. and Lonsdale, J. (1992) *Unhappy Valley: Conflict in Kenya and Africa*, 2 vols, London: James Currey.

Berry, S. (1985) *Fathers Work for Their Sons: Accumulation, Mobility and Class Formation in an Extended Yoruba Community*, Berkeley, CA and London: University of California Press.

—— (1993) *No Condition is Permanent: The Social Dynamics of Agrarian Change in Sub-Saharan Africa*, Madison, WI: University of Wisconsin Press.

Bigsten, A. (1978) *Regional Inequality and Development: A Case Study of Kenya*, Goteborg: University of Goteborg.

Bigsten, A. and Kayizzi-Mugera, S. (1995) 'Rural Sector Response to Economic Crisis in Uganda', *Journal of International Development* 6(4): 181–210.

Binswanger, H.P. and Deininger, K. (1993) 'South African Land Policy – The Legacy of History and Current Options', *World Development* 21(9): 1451–75.

Bingswanger, H.P. and Rosenzweig, M. (1986) 'Behavioural and Material Determinants of Production Relations in Agriculture', *Journal of Development Studies* 22(3): 503–39.

Bloom, D. and Sachs, J. (1998) 'Geography, Demography and Economic Growth in Africa', *Brookings Papers on Economic Activity*, 2: 207–95.

Booth, D. *et al.* (1993) *Social, Economic and Cultural Change in Contemporary Tanzania: A People-oriented Focus*, Stockholm: SIDA, report to SIDA, commissioned through the Development Studies Unit, Department of Social Anthropology, Stockholm University.

Bophuthatswana, Republic of (1993) *1991 Population Census*, Mmabatho: Statistics Branch, Department of Economics, Energy Affairs, Mines and Planning.

Boserup, E. (1970) *Woman's Role in Economic Development*, London: Allen and Unwin.

Bowie, F., Kirkwood, D. and Ardener, S. (1993) *Women and Missions: Past and Present*, London: Berg.

Bozzoli, B. (1983) 'Marxism, Feminism and South African Studies', *Journal of Southern African Studies* 9(2): 139–71.

Brett, E.A. (1973) *Colonialism and Underdevelopment in East Africa: The Politics of Economic Change 1919–39*, London: Heinemann.

Breutz, P.-L. (1955) *The Tribes of Mafeking District*, Department of Native Affairs, Ethnological Publications No. 32, Pretoria: Government Printer.

Bryceson, D. (1996) 'Deagrarianization and Rural Employment in Sub-Saharan Africa: A Sectoral Perspective', *World Development* 24(1): 97–111.

—— (1999) 'Sub-Saharan Africa Betwixt and Between: Rural Livelihood Practices and Policies', Leiden: African Studies Centre, Working Paper No. 43.

Bülow, D. von (1992) 'Bigger than Men? Gender Relations and their Changing Meaning in Kipsigis Society, Kenya', *Africa* 62(4): 522–46.

Bülow, D. von and Sørensen, A. (1988) *Contract Farming – Does it Benefit Women?*, Copenhagen: Centre for Development Research, Working Paper 88.5.

Bureau of Market Research, University of South Africa (1989) 'A Study Undertaken on Behalf of the Department of Economic Affairs of the Republic of Bophuthatswana', 89/14.

Butterman, J.M. (1979) 'Luo Social Formations in Change: Karachuonyo and Kanyamkago, c.1800–1945', Ph.D. dissertation, Syracuse University.

Carney, D. (1998) 'Implementing the Sustainable Rural Livelihoods Approach', in D. Carney (ed.), *Sustainable Rural Livelihoods: What Contribution Can We Make?*, London: Department for International Development.

Carter, M.R. (1997) 'Environment, Technology and the Social Articulation of Risk in West African Agriculture', *Economic Development and Cultural Change* 45(3): 557–91.

Central Settlement Planning Team (1996) 'Central District Perspective', prepared for the Central District Council, for the RDP Planning Office, North West Province, Mafikeng: SETPLAN, Report Ref. 3222/R2.

Chambers, R., Longhurst, R. and Pacey, A. (eds) (1981) *Seasonal Dimensions to Rural Poverty*, London: Frances Pinter.

Chant, S. (1997) *Women-Headed Households: Diversity and Dynamics in the Developing World*, Basingstoke: Macmillan.

Chayanov, A.V. (1966) *The Theory of Peasant Economy*, trans. D. Thorner *et al.*, Homewood, IL: Richard D. Irwin.

Christensen, G. and Stack, J. (1992) 'The Dimensions of Household Food Insecurity in Zimbabwe, 1980–1991', Food Studies Group Working Paper No. 5, Oxford.

Chuta, E. and Liedholm, C. (1990) 'Rural Small-scale Industry: Empirical Evidence and Policy Issues', in C. Eicher and J. Staatz (eds), *Agricultural Development in the Third World*, Baltimore, MD: Johns Hopkins University Press.

Clapp, R. (1994) 'The Moral Economy of the Contract', in P. Little and M. Watts (eds), *Living Under Contract: Contract Farming and Agrarian Transformation in Sub-Saharan Africa*, Madison, WI: University of Wisconsin Press.

Clayton, A. and Savage, D. (1974) *Government and Labour in Kenya: 1895–1963*, London: Frank Cass.

Cliffe, L. (1978) 'Labour Migration and Peasant Differentiation: the Zambian Case', *Journal of Peasant Studies* 5(3): 326–46.

Cliffe, L. *et al.* (eds) (1975) *Rural Co-operation in Tanzania*, Dar es Salaam: Tanzania Publishing House.

Cohen, A. (1969) *Custom and Politics in Urban Africa: A Study of Hausa Migrants in Yoruba Towns*, London: Routledge and Kegan Paul.

Cohen, D.W. and Atieno Odhiambo, E.S. (1989) *Siaya: The Historical Anthropology of an African Landscape*, London: James Currey.

—— (1992) *Burying SM: The Politics of Knowledge and the Sociology of Power in Africa*, Portsmouth, NH: Heinemann, London: James Currey.

Collier, P. and Lal, D. (1986) *Labour and Poverty in Kenya 1900–1980*, Oxford: Clarendon Press.

Collier, P. and Gunning, J.W. (1999) 'Explaining African Economic Performance', *Journal of Economic Literature* XXXVII: 64–111.

Comaroff, J. and Comaroff, J. (1997) *Of Revelation and Revolution: The Dialectics of Modernity on a South African Frontier*, vol. 2, Chicago, IL: University of Chicago Press.

Coplan, D. (1993) ' "Damned if we know." Public Policy, Labour Law and the Future of the Migrant Labour System', paper presented to the Conference of Labour Law, Durban.

Corbett, J. (1994) 'Livelihoods, Food Security and Nutrition in a Drought Prone Part of Zimbabwe', Department for International Development, ESCOR Project R4685, final report.

Coulson, A. (1982) *Tanzania: A Political Economy*, Oxford: Clarendon Press.

Cowen, M. (1975) 'Wattle Production in the Central Province: Capital and Household Commodity Production, 1903–1964', University of Nairobi, Institute for Development Studies, mimeo.

—— (1981a) 'Commodity Production in Kenya's Central Province', in J. Heyer *et al.* (eds), *Rural Development in Tropical Africa*, London: Macmillan.

—— (1981b) 'The Agrarian Problem: Notes on the Nairobi Discussion', *Review of African Political Economy* 20: 57–73.

Crehan, K. (1997) *The Fractured Community: Landscapes of Power and Gender in Rural Zambia*, Berkeley, CA: University of California Press.

Crush, J. (1995) 'Mine Migrancy in the Contemporary Era', in J. Crush and W. James (eds), *Crossing Boundaries: Mine Migrancy in a Democratic South Africa*, IDASA/IDRC.

Davenport, R. (1987) 'Can Sacred Cows be Culled? A Historical Review of Land Policy in South Africa, With Some Questions About the Future', *Development Southern Africa* 4(3).

Davison, J. (1988) 'Who Owns What? Land Registration and Tensions in Gender Relations of Production in Kenya', in J. Davison (ed.), *Agriculture, Women and Land: The African Experience*, Boulder, CO: Westview.

Delius, P. (1996) *A Lion Amongst the Cattle: Reconstruction and Resistance in the Northern Transvaal*, Portsmouth, NH: Heinemann, Oxford: James Currey.

De Wilde, J. (1967) *Experiences with Agricultural Development in Tropical Africa*, vol. 2, Baltimore, MD: Johns Hopkins Press.

Donge, J.K. van (1992) 'Agricultural Decline in Tanzania: The Case of the Uluguru Mountains', *African Affairs* 91: 73–94.

Drummond, J. (1990) 'Rural Land Use and Agricultural Production in Dinokana Village, Bophuthatswana', *GeoJournal* 22.

—— (1995) 'Development and Change: Irrigation and Agricultural Production in Dinokana Village, North West Province, South Africa', in T. Binns (ed.), *People and Environment in Africa*, Chichester, UK: John Wiley.

Drummond, J. and Manson, A. (1993) 'The Rise and Demise of African Agricultural Production in Dinokana Village, Bophuthatswana', *Canadian Journal of African Studies* 27(3): 462–79.

DuPré, C. (1968) *The Luo of Kenya: An Annotated Bibliography*, Washington, DC: Institute for Cross-Cultural Research.

Ellis, F. (1998) 'Survey Article: Household Strategies and Rural Livelihood Diversification', *Journal of Development Studies* 35(1): 1–38.

—— (forthcoming) *Rural Livelihood Diversity in Developing Countries: analysis, policy, methods*, London: Department for International Development, ESCOR, Project No. R6646.

Emmett, A.B., van Zyl, J.A. and Kok, P.C. (1988) 'Findings from a Survey to Establish Baseline Data for a Population Development Programme in Bophuthatswana, 1988',

Memorandum to the Department of Population Development, Republic of Bophuthatswana, Human Sciences Research Council.

Evans, A. (1993) ' "Contracted Out": Some Reflections on Gender, Power and Agrarian Institutions', *IDS Bulletin* 24(3): 21–30.

Evans-Pritchard, E.E. (1949) 'Luo Tribes and Clans', *Rhodes-Livingstone Journal* 7: 24–40; reprinted in E.E. Evans-Pritchard, *The Position of Women in Primitive Societies and Other Essays in Social Anthropology*, London: Faber and Faber, 1965.

Fearn, H. (1961) *An African Economy: a study of the economic development of the Nyanza Province of Kenya, 1903–1953*, London: Oxford University Press.

Ferguson, J. (1992) 'The Cultural Topography of Wealth – Commodity Paths and the Structure of Property in Rural Lesotho', *American Anthropologist* 94(1): 55–73.

Folbre, N. (1986) 'Hearts and Spades – Paradigms of Household Economics', *World Development* 14(2): 245–55.

—— (1994) *Who Pays for the Kids? Gender and Structures of Constraint*, London: Routledge.

Francis, E. (1999) 'Rural Livelihoods in Madibogo: A Preliminary Report', University of Manchester, Institute for Development Policy and Management, Multiple Livelihoods and Social Change Project Working Paper No. 6.

Francis, E. and Hoddinott, J. (1993) 'Migration and Differentiation in Western Kenya: A Tale of Two Sub-locations', *Journal of Development Studies* 30(1): 115–45.

Francis, E. and Williams, G. (1993) 'The Land Question', *Canadian Journal of African Studies* 27(3): 380–403.

Freyhold, von M. (1972) 'The Workers, the Nizers and the Peasants', mimeo, Dar es Salaam: Department of Sociology, University of Dar es Salaam.

—— (1979) *Ujamaa Villages in Tanzania: Analysis of a Social Experiment*, London: Heinemann.

Furedi, F. (1989) *The Mau Mau War in Perspective*, London: James Currey, Nairobi: Heinemann Kenya and Athens, OH: Ohio University Press.

Gaitskell, D. (1990) 'Devout Domesticity? A Century of African Women's Christianity in South Africa', in C. Walker (ed.), *Women and Gender in Southern Africa to 1945*, Cape Town: David Philip.

Gluckman, M. (1950) 'Kinship and Marriage Among the Lozi of Northern Rhodesia and the Zulu of Natal', in A.R. Radcliffe-Brown and D. Forde (eds), *African Systems of Kinship and Marriage*, London: Oxford University Press for the International African Institute (reprinted London: Kegan Paul International, 1987).

Goldenberg, D.A. (1982) ' "We are all Brothers": The Suppression of Consciousness of Socio-economic Differentiation in a Kenya Luo lineage', Ph.D. dissertation, Brown University.

Grillo, R.D. (1973) *African Railwaymen: Solidarity and Opposition in an East African Labour Force*, Cambridge: Cambridge University Press.

Grown, C.A. and Sebstad, J. (1989) 'Introduction: Toward a Wider Perspective on Women's Employment', *World Development* 17(7): 937–52.

Guy, J. (1990) 'Gender Oppression in South Africa's Precapitalist Societies', in C. Walker (ed.), *Women and Gender in Southern Africa to 1945*, Cape Town: David Philip and London: James Currey.

Guyer, J. (1997) *An African Niche Economy: Farming to Feed Ibadan 1968–88*, Edinburgh: Edinburgh University Press, for the International African Institute.

Haggblade S., Hazell, P. and Brown, J. (1989) 'Farm-nonfarm Linkages in Rural Sub-Saharan Africa', *World Development* 17(8): 1173–201.

Harriss, J. (ed.) (1982) *Rural Development: Theories of Peasant Economy and Agrarian Change*, London: Hutchinson.

Hart, G. (1995) 'Gender and Household Dynamics: Recent Theories and their Implications', in M. Quibria (ed.), *Critical Issues in Asian Development*, Hong Kong: Oxford University Press.

Hart, K. (1973) 'Informal Income Opportunities and Urban Employment in Ghana', *Journal of Modern African Studies* 11(1): 61–89.

—— (1992) 'Market and State after the Cold War: The Informal Economy Reconsidered', in R. Dilley (ed.), *Contesting Markets*, Edinburgh: Edinburgh University Press.

Haugerud, A. (1981) 'Economic Differentiation Among Peasant Household: A Comparison of Embu Coffee and Cotton Zones', University of Nairobi, Institute for Development Studies, Working Paper No. 383.

—— (1995) *The Culture of Politics in Modern Kenya*, Cambridge: Cambridge University Press.

Hay, M. (1972) 'Economic Change in Luoland: Kowe, 1890–1945', Ph.D. dissertation, Madison: University of Wisconsin, Department of History.

—— (1976) 'Luo Women and Economic Change During the Colonial Period', in N. Hafkin and E. Bay (eds), *Women in Africa: Studies in Social and Economic Change*, Stanford, CA: Stanford University Press.

—— (1982) 'Women as Owners, Occupants and Managers of Property in Colonial Western Kenya', in M. Hay and M. Wright (eds), *African Women and the Law: Historical Perspectives*, Boston, MA: Boston University African Studies Center.

Heald, S. (1991) 'Tobacco, Time and the Household Economy in Two Kenyan Societies: The Teso and the Kuria', *Comparative Studies in Society and History* 33(1): 130–57.

Heyer, J. (1967) 'The Economics of Small-Scale Farming in Lowland Machakos', University of Nairobi, Institute for Development Studies, Occasional Paper No. 1.

—— (1981) 'Agricultural Development Policy in Kenya from the Colonial Period to 1975', in J. Heyer, P. Roberts and G. Williams (eds), *Rural Development in Tropical Africa*, London: Macmillan.

—— (1996) 'The Complexities of Rural Poverty in Sub-Saharan Africa', *Oxford Development Studies*, 24(3): 281–97.

Heyer, J., Maitha, J. and Senga, W. (eds) (1976) *Agricultural Development in Kenya: An Economic Assessment*, Nairobi: Oxford University Press.

Heyer, J. and Waweru, J. (1976) 'The Development of the Small Farm Areas', in J. Heyer, J. Maitha and W. Senga (eds), *Agricultural Development in Kenya: An Economic Assessment*, Nairobi: Oxford University Press.

Hinga, S.N. and Heyer, J. (1976) 'The Development of Large Farms', in J. Heyer, J. Maitha and W. Senga (eds), *Agricultural Development in Kenya: An Economic Assessment*, Nairobi: Oxford University Press.

Hoddinott, J. (1989) 'Migration, Accumulation and Old Age Security in Western Kenya', D.Phil. thesis, University of Oxford.

Hulme, D. and Mosley, P. (1996) *Finance Against Poverty*, vol. 1, London: Routledge.

Hyden, G. (1980) *Beyond Ujamaa in Tanzania: Underdevelopment and an Uncaptured Peasantry*, London: Heinemann.

Ingle, C. (1972) *From Village to State in Tanzania: The Politics of Rural Development*, Ithaca, NY: Cornell University Press.

International Labour Organisation (1972) *Employment, Incomes and Equality: A Strategy for Increasing Productive Employment in Kenya*, Geneva: International Labour Office.

Jackson, C. (nd.) 'Conjugal Contracts and Agency in Rural Zimbabwe Environments', manuscript.

—— (1997) 'Working Bodies and Gender Divisions of Labour', manuscript.

Jackson, J. and Cheater, A. (1994) 'Contract Farming in Zimbabwe: Case Studies of Sugar, Tea and Coffee', in P. Little and M. Watts (eds), *Living Under Contract: Contract Farming and Agrarian Transformation in Sub-Saharan Africa*, Madison, WI: University of Wisconsin Press.

Jackson, J. and Collier, P. (1988) 'Incomes, Poverty and Food Security in the Communal Lands of Zimbabwe', University of Harare, Department of Rural and Urban Planning, Occasional Paper.

Jaffee, S. (1994) 'Contract Farming in the Shadow of Competitive Markets: The Experience of Kenyan Horticulture', in P. Little and M. Watts (eds), *Living Under Contract: Contract Farming and Agrarian Transformation in Sub-Saharan Africa*, Madison, WI: University of Wisconsin Press.

Jamal, V. and Weeks, J. (1993) *Africa Misunderstood, or Whatever Happened to the Rural-Urban Gap?*, London: Macmillan.

Jambiya, G. (1998) 'The Dynamics of Population, Land Scarcity, Agriculture and Non-Agricultural Activities: West Usambara Mountains, Lushoto District, Tanzania', Dar es Salaam, Institute of Resource Assessment and Leiden, African Studies Centre Working Paper No. 28.

Jeppe, W. (1980) *Bophuthatswana Land Tenure and Development*, Cape Town: Maskew Miller.

Johnson, G.E. and Whitelaw, W.E. (1974) 'Urban-rural Income Transfers in Kenya: An Estimated-remittances Function', *Economic Development and Cultural Change* 22(3): 473–9.

Kabeer, N. (1991) 'Gender, Production and Well-Being: Rethinking the Household Economy', Institute for Development Studies, Sussex, Discussion Paper no. 288.

—— (1994) *Reversed Realities: Gender Hierarchies in Development Thought*, London: Verso.

Kandiyoti, D. (1988) 'Bargaining With Patriarchy', *Gender and Society* 2(3): 274–90.

—— (1998) 'Gender, Power and Contestation: "Bargaining with Patriarchy" revisited', in C. Jackson and R. Pearson (eds), *Feminist Visions of Development*, London: Routledge.

Kanogo, T. (1987) *Squatters and the Roots of Mau Mau: 1905–63*, London: James Currey, Nairobi: Heinemann Kenya and Athens, OH: Ohio University Press.

Keenan, J. (1984) 'The Penetration of Agricultural Capital in South Africa's Bantustans', *South African Review* 2.

Kenya, Colony and Protectorate of (1937) *Annual Report*, Education Department.

—— (1954) *Annual Report*, Education Department, Nairobi: Government Printer.

Kenya, Republic of (1977) *Annual Report*, Ministry of Education.

—— (1986) *Statistical Abstract 1986*, Central Bureau of Statistics, Ministry of Planning and National Development, Nairobi: Government Printer.

King, K. (1996) *Jua Kali Kenya: Change and Development in an Informal Economy, 1970–95*, London: James Currey.

Kinsey, B., Burger, K. and Gunning, J. (1998) 'Coping with Drought in Zimbabwe: Survey Evidence on Household Responses to Risk', *World Development* 26: 207–27.

Kitching, G. (1980) *Class and Economic Change in Kenya: The Making of an African Petite Bourgeoisie*, New Haven, CT: Yale.

Klerk, M. de (1984) 'Seasons that Will Never Return: The Impact of Farm Mechanisation on Employment, Incomes and Population Distribution in the Western Transvaal', *Journal of Southern African Studies* 11(1): 84–105.

—— (1996) 'A Qualifying Perspective', in M. Lipton, M. de Klerk and M. Lipton (eds), *Land, Labour and Livelihoods in Rural South Africa*, vol. 1, Durban: Indicator Press.

Kotzé, J. (1992) 'Children and Family in Rural Settlement in Gazankulu', *African Studies* 51(2): 143–66.

Lawrence, M. and Manson, A. (1994) ' "The Dog of the Boers": The Rise and Fall of Mangope in Bophuthatswana', *Journal of Southern African Studies* 20(3): 447–61.

Leach, M. and Mearns, R. (eds) (1996) *The Lie of the Land: Challenging Received Wisdom on the African Environment*, Oxford: James Currey and Portsmouth, NH: Heinemann.

Leo, C. (1984) *Land and Class in Kenya*, Toronto, Ont.: University of Toronto Press.

Levin, R. and Weiner, D. (eds) (1997) *'No More Tears': Struggles for Land in Mpumalanga, South Africa*, Trenton, NJ: Africa World Press.

Leys, C. (1975) *Underdevelopment in Kenya: The Political Economy of Neo-colonialism*, London: Heinemann.

—— (1994) 'Confronting the African Tragedy', *New Left Review* 204: 33–47.

Liedholm, C. and Mead, D. (1993) 'The Structure and Growth of Microenterprises in Southern and Eastern Africa: Evidence from Recent Surveys', Occasional Paper no. 1, GEMINI Project, Working Paper No. 6, Development Alternatives, Inc., Bethesda, Maryland.

Lipton, M. (1996) 'Rural Reforms and Rural Livelihoods: The Contexts of International Experience', in M. Lipton, M. de Klerk and M. Lipton (eds), *Land, Labour and Livelihoods in Rural South Africa*, vol. 1, Durban: Indicator Press.

Lipton, M., de Klerk, M. and Lipton, M. (eds) (1996a) *Land, Labour and Livelihoods in Rural South Africa*, vol. 1, Durban: Indicator Press.

Lipton, M., Ellis, F. and Lipton, M. (eds) (1996b) *Land, Labour and Livelihoods in Rural South Africa*, vol. 2, Durban: Indicator Press .

Lipton, M. and Ravallion, M. (1995) 'Poverty and Policy', in J. Behrman and T.N. Srinivasan (eds), *Handbook of Development Economics*, Vo. IIIB, Amsterdam: Elsevier.

Little, P. (1994) 'Contract Farming and the Development Question', in P. Little and M. Watts (eds), *Living Under Contract: Contract Farming and Agrarian Transformation in Sub-Saharan Africa*, Madison, WI: University of Wisconsin Press.

Livingstone, I. (1986) *Rural Development, Employment and Incomes in Kenya*, Aldershot: Gower, for ILO: JASPA.

Lonsdale, J. (1964) 'A Political History of Nyanza: 1883–1945', Ph.D. thesis, University of Cambridge.

Low, A. (1986) *Agricultural Development in Southern Africa: Farm Household Theory and the Food Crisis*, London: James Currey.

Loxley, J. (1990) 'Structural Adjustment in Africa: Reflections on Ghana and Zambia', *Review of African Political Economy* 47: 8–27.

—— (1995) 'Rural Labour Markets in an Adjusting Mineral Economy: Zambia', in V. Jamal (ed.), *Structural Adjustment and Rural Labour Markets in Africa*, Basingstoke: Macmillan.

McAllister, P. (1989) 'Resistance to "Betterment" in the Transkei: A Case Study from Willowvale District', *Journal of Southern African Studies* 15(2): 346–68.

MacGaffey, J. *et al.* (1991) *The Real Economy of Zaire*, London: James Currey and Philadelphia, PA: University of Pennsylvania Press.

McIntosh, A. and Vaughan, A. (1996) 'Enhancing Rural Livelihoods in South Africa: Myths and Realities', in Lipton *et al.* (1996b).

Mackenzie, F. (1993) '"A Piece of Land Never Shrinks": Reconceptualizing Land Tenure in a Smallholding District, Kenya', in T. Bassett and D. Crummey (eds), *Land in African Agrarian Systems*, Madison, WI: University of Wisconsin Press.

—— (1998) *Land, Ecology and Resistance in Kenya, 1880–1952*, Edinburgh: Edinburgh University Press for the International African Institute.

Madulu, N.F. (1998) 'Changing Lifestyles in Farming Societies of Sukumaland: Kwimba District, Tanzania', Dar es Salaam, Institute of Resource Assessment and Leiden, African Studies Centre Working Paper no. 33.

Maher, C. (1937) 'Soil Erosion and Land Utilisation in the Ukamba Reserve (Machakos)', report to the Department of Agriculture, Mss. Afr. S.545, Rhodes House Library, Oxford.

Mamdani, M. (1996) *Citizen and Subject: Contemporary Africa and the Legacy of Late Colonialism*, Kampala: Fountain Publishers, Cape Town: David Philip, London: James Currey.

Marcus, T. (1989) *Modernising Super-Exploitation: Restructuring South African Agriculture*, London: Zed Books.

Market Research Ltd (1965) *Dar es Salaam Social Survey*, in A. Tripp (1997) *Changing the Rules: The Politics of Liberalization and the Urban Informal Economy in Tanzania*, Berkeley and Los Angeles, CA: University of California Press.

Marris, P. and Somerset, A. (1971) *African Businessman: A Study of Entrepreneurship and Development in Kenya*, London: Routledge and Kegan Paul.

Mather, C. and Drummond, J. (nd.) 'Agricultural Restructuring and the Post-Apartheid Agricultural Regime', typescript, J. Drummond, Department of Geography, University of the North West, Mafikeng.

Meyers, L. R. (1982) 'Socioeconomic determinants of credit adoption in a semiarid district of Kenya' Ph.D. dissertation, Cornell University.

Moock, J. (1976) 'The Migration Process and Differential Economic Behaviour in South Maragoli, Western Kenya', Ph.D. dissertation, Columbia University.

Moore, H. (1994) 'Households and Gender in a South African Bantustan', *African Studies* 53(1): 137–42.

Moore, H. and Vaughan, M. (1994) *Cutting Down Trees: Gender, Nutrition and Agricultural Change in the Northern Province of Zambia, 1890–1990*, Portsmouth, NH: Heinemann, London: James Currey, Lusaka: University of Zambia Press.

Morrissy, J.D. (1974) *Agricultural Modernization through Production Contracting: The Role of the Fruit and Vegetable Processor in Mexico and Central America*, New York: Praeger.

Mukras, M.S., Oucho, J.O. and Bamberger, M. (1985) 'Resource Mobilization and the Household Economy in Kenya', *Canadian Journal of African Studies* 19(2): 409–21.

Murray, C. (1981) *Families Divided: The Impact of Migrant Labour in Lesotho*, Cambridge: Cambridge University Press.

—— (1992) *Black Mountain: Land, Class and Power in the Eastern Orange Free State, 1880s–1980s*, Edinburgh: Edinburgh University Press for the International African Institute.

—— (1995) 'Structural Unemployment, Small Towns and Agrarian Change in Rural South Africa', *African Affairs* 94: 5–22.

—— (1998) 'Changing Livelihoods in Qwaqwa: Research Questions and Methods of Study', University of Manchester, Institute for Development Policy and Management, Multiple Livelihoods and Social Change Project Working Paper No. 1.

Murray, C. and Williams, G. (1994) 'Editorial: Land and Freedom in South Africa', *Review of African Political Economy*, 61: 315–24.

Närman, A. (1995) 'The Dilemmas Facing Kenya School Leavers: Surviving in the City or a Force for Local Mobilization?', in J. Baker and T.A. Aina (eds), *The Migration Experience in Africa*, Uppsala: Nordiska Afrikainstitutet.

Niehaus, I. (1994) 'Disharmonious Spouses and Harmonious Siblings', *African Studies* 53(1): 115–35.

Nyangira, N. (1975) *Relative Modernization and Public Resource Allocation in Kenya, a Comparative Analysis*, Nairobi: East African Literature Bureau.

Oboler, R.S. (1994) 'The House-Property Complex and African Social Organisation', *Africa* 64(3): 342–58.

Ockwell, A.P., Parton, K.A., Nguw, S. and Muhammad, L. (1990) 'Relationship Between the Farm Household and Adoption of Improved Practices in the Semi-Arid Tropics of Eastern Kenya', *Dryland Farming Symposium: a Search for Strategies for Sustainable Dryland Cropping in Semi-Arid Eastern Kenya*, held in conjunction with the review of KARI/ACIAR/CSIRO Dryland Project, Nairobi, December 1990.

Odaga, A. (1990) '"Kech en Mar Pesa": Gender and Livelihood in a Western Kenya Sub-location', D.Phil. thesis, University of Oxford.

Ogot, B. (1967) *History of the Southern Luo, Volume 1: Migration and Settlement 1500–1900*, Nairobi: East Africa Publishing House.

Ojwang, J.B. and Mugambi, J.N.K. (eds) (1989) *The S.M. Otieno Case: Death and Burial in Modern Kenya*, Nairobi: Nairobi University Press.

Ominde, S.H. (1968) *Land and Population Movements in Kenya*, London: Heinemann.

Orvis, S. (1985) 'A Patriarchy Transformed: Reproducing Labour and the Viability of Smallholder Agriculture in Kisii', Institute for Development Studies, Working Paper No. 434.

Pala, A.O. (1977) 'Changes in Economy and Ideology: a Study of the Joluo of Kenya (With Special Reference to Women)', Ph.D. dissertation, Harvard University.

Pankhurst, D. (1991) 'Constraints and Incentives in "Successful" Zimbabwean Peasant Agriculture: The Interaction Between Gender and Class', *Journal of Southern African Studies* 17(4): 611–32.

Pankhurst, D. and Jacobs, S. (1988) 'Land Tenure, Gender Relations and Agricultural Production: The Case of Zimbabwe's Peasantry', in J. Davison (ed.), *Women and Land in Africa*, Boulder, CO: Westview.

Parkin, D. (1975) 'Migration, Settlement and the Politics of Unemployment: A Nairobi Case Study', in D. Parkin (ed.), *Town and Country in Central and Eastern Africa*, London: Oxford University Press.

—— (1978) *The Cultural Definition of Political Response: Lineal Destiny Among the Luo*, London: Academic Press.

Peperzak, P. (1985) 'IFC Plays Pivotal Role in Supporting Agriculture: Project Criteria Outlined', *Agribusiness Worldwide* Nov–Dec: 12–21.

Peters, P. (1983) 'Gender, Development Cycles and Historical Process: A Critique of Recent Research on Women in Botswana', *Journal of Southern African Studies* 10(1): 100–22.

Platzky, L. and Walker, C. (1985) *The Surplus People: Forced Removals in South Africa*, Johannesburg: Ravan Press.

Portes, A. and Walton, J. (1981) *Labor, Class and the International System*, New York: Academic Press.

Potash, B. (1978) 'Some Aspects of Marital Stability in a Rural Luo Community', *Africa* 48(4): 380–96.

Pottier, J. (1988) *Migrants No More: Settlement and Survival in Mambwe Villages, Zambia*, Manchester: Manchester University Press.

Potts, D. (1995) 'Shall We Go Home? Increasing Urban Poverty in African Cities and Migration Processes', *Geographical Journal* 161(3): 676–98.

—— (1997) 'Urban Lives: Adopting New Strategies and Adapting Rural Links', in C. Rakodi (ed.), *The Urban Challenge in Africa: Growth and Management of Its Large Cities*, Tokyo: United Nations University Press.

—— (1999) 'Urban Employment, Unemployment and Migrants in Africa: Evidence from Harare 1985–1994', paper presented at Development Studies Seminar, University of Manchester, 2 November.

Potts, D. and Mutambirwa, C.C. (1990) 'Rural-Urban Linkages in Contemporary Harare: Why Migrants Need Their Land', *Journal of Southern African Studies* 16(4): 245–64.

Putzel, J. (1997) 'Accounting for the "Dark Side" of Social Capital: Reading Robert Putnam on Democracy', *Journal of International Development* 9(7): 939–49.

Rahman, H.Z. and Hossain, M. (eds) (1992) 'Re-thinking Rural Poverty: A Case for Bangladesh', Analysis of Poverty Trends Project, Dhaka: Bangladesh Institute of Development Studies, draft.

Raikes, P. (1993) 'Business as Usual: National and Local Food Marketing in Kenya', Copenhagen: Centre for Development Research, Working Paper 93.9.

Rakodi, C. (1991) 'Women's Work or Household Strategies?', *Environment and Urbanization* 3(2): 39.

Reardon, T. (1997) 'Using Evidence of Household Income Diversification to Inform Study of the Rural Nonfarm Labor Market in Africa', *World Development* 25(5): 735–47.

Reimer, K. (1987) 'The Impact of Development on the Sexual Division of Labour: A Botswana Case Study', MA dissertation, University of the Witwatersrand.

Rempel, H. and Lobdell, R. (1978) 'The Role of Urban-To-Rural Remittances in Rural Development', *Journal of Development Studies* 14(3): 324–34.

Richards, A. (1939) *Land, Labour and Diet in Northern Rhodesia: An Economic Study of the Bemba Tribe*, Oxford: Oxford University Press.

Richards, P. (1985) *Indigenous Agricultural Revolution: Ecology and Food Production in West Africa*, London: Hutchinson.

Roe, E. (1991) ' "Development Narratives", or Making the Best of Blueprint Development', *World Development* 19(4): 287–300.

Roodt, J. (1985) 'Capital Penetration and Rural Development in Bophuthatswana: A Case Study of the Ditsobotla Dryland Projects', MA dissertation, University of the Witwatersrand.

—— (1988) 'Bophuthatswana's State Farming Projects: Is Failure Inevitable?', in C. Cross and R. Haines (eds), *Towards Freehold: Options for Land and Development in South Africa's Black Rural Areas*, Cape Town: Juta Press.

Ross, M. and Weisner, T. (1977) 'The Rural-Urban Migrant Network in Kenya: Some General Implications', *American Ethnologist* 4(2): 359–75.

Rukandema, M., Mavua, J.K. and Audi, P.O. (1981) *The Farming System of Lowland Machakos, Kenya: Farm Survey Results from Mwala*, Farming Systems Economic Research Programme Technical Report (Kenya), No. 1, Nairobi: Ministry of Agriculture.

Sahn, D. (1994) 'The Impact of Macroeconomic Adjustment on Incomes, Health and Nutrition: Sub-Saharan Africa in the 1980s', in G. Cornia and G. Helleiner (eds), *From Adjustment to Development in Africa: Conflict, Controversy, Convergence, Consensus?*, London: Macmillan.

Satterthwaite, D. (1996) *The Scale and Nature of Urban Change in the South*, London: IIED.

Schatzberg, M. (ed.) (1987) *The Political Economy of Kenya*, New York and London: Praeger.

Schmidt, E. (1992) *Peasants, Traders and Wives: Shona Women in the History of Zimbabwe, 1870–1939*, Portsmouth, NH: Heinemann, Harare: Baobab, London: James Currey.

Scoones, I. (1998) 'Sustainable Rural Livelihoods: A Framework for Analysis', Brighton: Institute for Development Studies, Working Paper No. 72.

Scoones, I. *et al.* (1996) *Hazards and Opportunities. Farming Livelihoods in Dryland Africa: Lessons from Zimbabwe*, London: Zed Press.

Scott, A. (1994) *Divisions and Solidarities: Gender, Class and Employment in Latin America*, London: Routledge.

Sen, A. (1990) 'Co-operative Conflicts', in I. Tinker (ed.), *Persistent Inequalities*, Oxford: Oxford University Press.

Sender, J. and Smith, S. (1986) *The Development of Capitalism in Africa*, London: Routledge.

—— (1990) *Poverty, Class and Gender in Rural Africa: a Tanzanian Case Study*, London: Methuen.

Seppälä, P. (1995) *The Changing Generation: The Devolution of Land Among the Babukusu in Western Kenya*, Helsinki: The Finnish Anthropological Society, Transaction No. 35.

—— (1996) 'The Politics of Economic Diversification: Reconceptualising the Rural Informal Sector in South-east Tanzania', *Development and Change* 27(3): 557–78.

Shanin, T. (ed.) (1988) *Peasants and Peasant Societies*, 2nd edn, London: Penguin.

Sharp, J. (1994) 'A World Turned Upside Down: Households and Differentiation in a South African Bantustan in the 1980s', *African Studies* 53(1): 71–88.

Sharp, J. and Spiegel, A. (1990) 'Women and Wages: Gender and Control of Income in Farm and Bantustan Households', *Journal of Southern African Studies* 16(3): 527–49.

Shipton, P. (1979) 'Two East African Systems of Land Rights', M.Litt. thesis, University of Oxford.

—— (1985) 'Land, Credit and Crop Transitions in Kenya', Ph.D. thesis, University of Cambridge.

—— (1988) 'The Kenyan Land Tenure Reform: Misunderstandings in the Public Creation of Private Property', in R.E. Downs and S.P. Reyna (eds), *Land and Society in Contemporary Africa*, Hanover, NH: University Press of New England.

—— (1989) *Bitter Money: Cultural Economy and Some African Meanings of Forbidden Commodities*, American Ethnological Society Monograph Series, no. 1.

Shirji, I. (1975) *Class struggles in Tanzania*, London: Heinemann and Dar es Salaam: Tanzania Publishing House.

Silberschmidt, M. (1992) 'Have Men become the Weaker Sex? Changing Life Situations in Kisii District, Kenya', *Journal of Modern African Studies* 30(2): 237–53.

Sørensen, A. and Bülow, D. von (1990) 'Gender and Contract Farming in Kericho, Kenya', Copenhagen: Centre for Development Research, Project Paper 90.4.

Southall, A. (1952) 'Lineage Formation among the Luo', *Memorandum* XXVI, International Africa Institute, London: Oxford University Press.

—— (1975) 'From Segmentary Lineage to Ethnic Association', in M. Owusu (ed.), *Colonialism and Change: Essays Presented to Lucy Mair*, The Hague: Mouton.

Spiegel, A. (1980) 'Rural Differentiation and the Diffusion of Migrant Labour Remittances in Lesotho', in P. Mayer (ed.), *Black Villagers in an Industrial Society: Anthropological Perspectives on Labour Migration in South Africa*, Cape Town: Oxford University Press.

Stacey, G. (1992) 'The Origin and Development of Commercial Farmers in the Ditsobotla and Molopo Regions of Bophuthatswana', M.Sc. (Agric.) dissertation, University of Pretoria.

Stacey, G., Van Zyl, J. and Kirsten, J. (1994) 'The Causes and Consequences of Agricultural Change: A Case Study of Agricultural Commercialization in the Communal Areas of South Africa', *Agrekon* 33(4).

Stichter, S. (1982) *Migrant Labour in Kenya: Capitalism and African Response, 1895–1975*, London: Longman.

Summers, R. and Heston, A. (1991) 'The Penn World Table (Mark 5): An Expanded Set of International Comparisons, 1950–1988', *Quarterly Journal of Economics* May: 327–68.

Sweetman, G. (1995) *The Miners Return: Changing Gender Relations in Lesotho's Ex-migrants' Families*, Norwich: University of East Anglia, GAID, 9.

Swynnerton, R.J.M. (1954) *A Plan to Intensify the Development of African Agriculture in Kenya*, Nairobi: Government Printer.

Throup, D. (1987) *Economic and Social Origins of Mau Mau: 1945–53*, London: James Currey, Nairobi: Heinemann Kenya and Athens, OH: Ohio University Press.

Throup, D. and Hornsby, C. (1997) *Multi-party Politics in Kenya: The Kenyatta and Moi States and the Triumph of the System in the 1992 Election*, London: James Currey.

Tiffen, M., Mortimore, M. and Gichuki, F. (1994) *More People, Less Erosion: Environmental Recovery in Kenya*, Chichester: John Wiley.

Tomlinson, F. (1955) *Summary of the Report of the Commission for the Socio-Economic Development of the Bantu Areas within the Union of South Africa*, Pretoria: Government Printer.

Tripp, A. (1997) *Changing the Rules: The Politics of Liberalization and the Urban Informal Economy in Tanzania*, Berkeley and Los Angeles, CA: University of California Press.

Turok, B. (ed.) (1979) *Development in Zambia*, London: Zed Press.

United Nations Development Programme (1999) *Human Development Report 1999*, New York: Human Development Report Office, United Nations Development Programme.

Vail, L. (1983) 'Political Economy of East-Central Africa', in D. Birmingham and P. Martin (eds), *A History of Central Africa*, vol. 2, London: Longman.

Van Zwanenberg, R.M.A. (1975) *Colonial Capitalism and Labour in Kenya: 1919–39*, Nairobi: East African Literature Bureau.

Vaughan, A. (1992) 'Options for Rural Restructuring', in R. Schrire (ed.), *Wealth or Poverty? Critical Choices for South Africa*, Cape Town: Oxford University Press.

Vaughan, M. (1983) 'Which Family? Problems in the Reconstruction of the History of the Family as an Economic and Cultural Unit', *Journal of African History* 24: 275–83.

Wasserman, G. (1976) *The Politics of Decolonization: Kenya Europeans and the Land Issue*, Cambridge: Cambridge University Press.

Watson, W. (1958) *Tribal Cohesion in a Money Economy: A Study of the Mambwe People of Northern Rhodesia*, Manchester: Manchester University Press.

Watts, M. (1994) 'Life Under Contract: Contract Farming, Agrarian Restructuring and Flexible Accumulation', in P. Little and M. Watts (eds), *Living Under Contract: Contract Farming and Agrarian Transformation in Sub-Saharan Africa*, Madison, WI: University of Wisconsin Press.

Wet, C. de (1989) 'Betterment Planning in a Rural Village in Keiskammahoek, Transkei', *Journal of Southern African Studies* 15(2): 326–45.

Whisson, M. (1964) *Change and Challenge: A Study of the Social and Economic Changes Among the Kenya Luo*, Nairobi: Christian Council of Kenya.

White, L. (1990) *The Comforts of Home: Prostitution in Colonial Nairobi*, Chicago, IL: University of Chicago Press.

Whitehead, A. (1984) ' "I'm Hungry, Mum": The Politics of Domestic Budgeting', in K. Young *et al.* (eds), *Of Marriage and the Market: Women's Subordination Internationally and its Lessons*, London: Routledge.

—— (1990) 'Food Crisis and Gender Conflict in the African Countryside', in H. Bernstein *et al.* (eds), *The Food Question*, London: Earthscan.

Williams, G. (1994) 'Land and Freedom', University of Natal, Durban: Centre for Social and Development Studies, Working Paper No. 12.

—— (1996) 'Setting the Agenda: A Critique of the World Bank's Rural Restructuring Program for South Africa', *Journal of Southern African Studies* 22(1): 139–66.

—— (1997) ' "Plus ça change, plus c'est la même chose": Land and Agricultural Policies in South Africa', paper presented to a conference on Sustainable Land Reform, Johannesburg.

Williams, S. and Karen, R. (1985) *Agribusiness and the Small Farmer: A Dynamic Partnership for Development*, Boulder, CO: Westview.

Wilson, G. (1961) *Luo Customary Law and Marriage Laws Customs* (sic), Nairobi: Government Printer.

World Bank (1993) *Options for Land Reform and Rural Restructuring in South Africa*, Johannesburg: Land and Agriculture Policy Centre.

World Bank (1999) *World Development Report 1998/99*, New York: Oxford University Press, for the World Bank.

Wright, C. (1993) 'Unemployment, Migration and Changing Gender Relations in Lesotho', Ph.D. thesis, University of Leeds.

Wright, M. (1983) 'Technology, Marriage and Women's Work in the History of Maize-growers in Mazabuka, Zambia: A Reconnaissance', *Journal of Southern African Studies* 10(1): 71–85.

Yawitch, J. (1981) *Betterment: The Myth of Homeland Agriculture*, Johannesburg: South African Institute of Race Relations.

Zinyama, L. (1995) 'Sustainability of Smallholder Food Production Systems in Southern Africa: The Case Zimbabwe', in T. Binns (ed.), *People and Environment in Africa*, Chichester: John Wiley.

Index